Lecture Notes in Computer Science 13781

More information about this series at https://link.springer.com/bookseries/558

Nicholas Olenev · Yuri Evtushenko ·
Milojica Jaćimović · Michael Khachay ·
Vlasta Malkova · Igor Pospelov (Eds.)

Optimization and Applications

13th International Conference, OPTIMA 2022
Petrovac, Montenegro, September 26–30, 2022
Revised Selected Papers

Springer

Editors
Nicholas Olenev (iD)
Dorodnitsyn Computing Centre
Moscow, Russia

Yuri Evtushenko (iD)
Dorodnitsyn Computing Centre
Moscow, Russia

Milojica Jaćimović (iD)
University of Montenegro
Podgorica, Montenegro

Michael Khachay (iD)
Krasovskii Institute of Mathematics
and Mechanics
Ekaterinburg, Russia

Vlasta Malkova (iD)
Dorodnitsyn Computing Centre
Moscow, Russia

Igor Pospelov (iD)
Dorodnitsyn Computing Centre
Moscow, Russia

ISSN 0302-9743   ISSN 1611-3349 (electronic)
Lecture Notes in Computer Science
ISBN 978-3-031-22542-0   ISBN 978-3-031-22543-7 (eBook)
https://doi.org/10.1007/978-3-031-22543-7

This Springer imprint is published by the registered company Springer Nature Switzerland AG
The registered company address is: Gewerbestrasse 11, 6330 Cham, Switzerland

Preface

This volume contains the second part of the refereed proceedings of the XIII International Conference on Optimization and Applications (OPTIMA 2022)[1].

Organized annually since 2009, the conference has attracted a significant number of researchers, academics, and specialists in many fields of optimization, operations research, optimal control, game theory, and their numerous applications in practical problems of data analysis and software development.

The broad scope of OPTIMA has made it an event where researchers involved in different domains of optimization theory and numerical methods, investigating continuous and discrete extremal problems, designing heuristics and algorithms with theoretical bounds, developing optimization software, and applying optimization techniques to highly relevant practical problems can meet together and discuss their approaches and results. We strongly believe that this facilitates collaboration between researchers working in optimization theory, methods, and applications, and those employing them to resolve valuable practical problems.

The conference was held during September 26–30, 2022, in Petrovac, Montenegro, in the picturesque Budvanian riviera on the azure Adriatic coast. For those who were not able to come to Montenegro this year, an online session was organized. The main organizers of the conference were the Montenegrin Academy of Sciences and Arts, Montenegro, the Dorodnicyn Computing Centre, FRC CSC RAS, Russia, and the University of Évora, Portugal. This year, the key topics of OPTIMA were grouped into seven tracks:

  (i)   Mathematical programming
  (ii) Global optimization
 (iii) Discrete and combinatorial optimization
 (iv) Optimal control
  (v) Optimization and data analysis
 (vi) Game theory and mathematical economics
(vii) Applications

The Program Committee (PC) and invited reviewers included more than one hundred well-known experts in continuous and discrete optimization, optimal control and game theory, data analysis, mathematical economy, and related areas from leading institutions of 25 countries: Argentina, Australia, Austria, Belgium, China, Finland, France, Germany, Greece, India, Israel, Italy, Lithuania, Kazakhstan, Mexico, Montenegro, the Netherlands, Poland, Portugal, Russia, Serbia, Sweden, Taiwan, the UK, and the USA. This year we received 70 submissions, mostly from Russia but also from Azerbaijan, Kazakhstan, Latvia, Montenegro, Poland, Portugal, and the USA. Each submission was reviewed in a single blind manner by at least three PC members or invited

---

[1] http://agora.guru.ru/display.php?conf=OPTIMA-2022.

reviewers, experts in their fields, to supply detailed and helpful comments. Out of 43 qualified submissions, the Program Committee decided to accept 17 papers to the first volume of the proceedings for publication in LNCS volume 13781. Thus the acceptance rate for the volume was about 40%.

In addition, after a short presentation of the candidate submissions, discussion at the conference, and subsequent revision, the Program Committee proposed 13 out of the remaining 26 papers to be included in this, second, volume of proceedings.

The conference featured two invited lecturers, plus several plenary and keynote talks. The invited lectures were as follows:

- Panos M. Pardalos, University of Florida, USA, "Computational Approaches for Solving Systems of Nonlinear Equations"
- Alexey Tret'yakov, Siedlce University of Natural Sciences and Humanities, Poland, "Degenerate Equality Constrained Optimization Problems and P-Regularity Theory"

We would like to thank all the authors for submitting their papers and the members of the PC for their efforts in providing exhaustive reviews. We would also like to express special gratitude to all the invited lecturers and plenary speakers.

October 2022

Nicholas Olenev
Yuri Evtushenko
Milojica Jaćimović
Michael Khachay
Vlasta Malkova
Igor Pospelov

Organization

Program Committee Chairs

Milojica Jaćimović	Montenegrin Academy of Sciences and Arts, Montenegro
Yuri G. Evtushenko	Dorodnicyn Computing Centre, FRC CSC RAS, Russia
Igor G. Pospelov	Dorodnicyn Computing Centre, FRC CSC RAS, Russia
Michael Yu. Khachay	Krasovsky Institute of Mathematics and Mechanics, Russia
Vlasta U. Malkova	Dorodnicyn Computing Centre, FRC CSC RAS, Russia
Nicholas N. Olenev	CEDIMES-Russie and Dorodnicyn Computing Centre, FRC CSC RAS, Russia

Program Committee

Majid Abbasov	St. Petersburg State University, Russia
Samir Adly	University of Limoges, France
Kamil Aida-Zade	Institute of Control Systems of ANAS, Azerbaijan
Alla Albu	Dorodnicyn Computing Centre, FRC CSC RAS, Russia
Alexander P. Afanasiev	Institute for Information Transmission Problems, RAS, Russia
Yedilkhan Amirgaliyev	Suleyman Demirel University, Kazakhstan
Anatoly S. Antipin	Dorodnicyn Computing Centre, FRC CSC RAS, Russia
Adil Bagirov	Federation University, Australia
Artem Baklanov	International Institute for Applied Systems Analysis, Austria
Evripidis Bampis	LIP6, Sorbonne Université, France
Olga Battaïa	ISAE-SUPAERO, France
Armen Beklaryan	National Research University Higher School of Economics, Russia
Vladimir Beresnev	Sobolev Institute of Mathematics, Russia
Anton Bondarev	Xi'an Jiaotong-Liverpool University, China
Sergiy Butenko	Texas A&M University, USA
Vladimir Bushenkov	University of Évora, Portugal

Igor A. Bykadorov	Sobolev Institute of Mathematics, Russia
Alexey Chernov	Moscow Institute of Physics and Technology, Russia
Duc-Cuong Dang	INESC TEC, Portugal
Tatjana Davidovic	Mathematical Institute of Serbian Academy of Sciences and Arts, Serbia
Stephan Dempe	TU Bergakademie Freiberg, Germany
Askhat Diveev	FRC CSC RAS and RUDN University, Russia
Alexandre Dolgui	IMT Atlantique, LS2N, CNRS, France
Olga Druzhinina	FRC CSC RAS, Russia
Anton Eremeev	Omsk Division of Sobolev Institute of Mathematics, SB RAS, Russia
Adil Erzin	Novosibirsk State University, Russia
Francisco Facchinei	Sapienza University of Rome, Italy
Vladimir Garanzha	Dorodnicyn Computing Centre, FRC CSC RAS, Russia
Alexander V. Gasnikov	National Research University Higher School of Economics, Russia
Manlio Gaudioso	Universita della Calabria, Italy
Alexander Yu. Gornov	Institute of System Dynamics and Control Theory, SB RAS, Russia
Edward Kh. Gimadi	Sobolev Institute of Mathematics, SB RAS, Russia
Andrei Gorchakov	Dorodnicyn Computing Centre, FRC CSC RAS, Russia
Alexander Grigoriev	Maastricht University, The Netherlands
Mikhail Gusev	N.N. Krasovskii Institute of Mathematics and Mechanics, Russia
Vladimir Jaćimović	University of Montenegro, Montenegro
Vyacheslav Kalashnikov	ITESM, Monterrey, Mexico
Maksat Kalimoldayev	Institute of Information and Computational Technologies, Kazakhstan
Valeriy Kalyagin	Higher School of Economics, Russia
Igor E. Kaporin	Dorodnicyn Computing Centre, FRC CSC RAS, Russia
Alexander Kazakov	Matrosov Institute for System Dynamics and Control Theory, SB RAS, Russia
Oleg V. Khamisov	L. A. Melentiev Energy Systems Institute, Russia
Andrey Kibzun	Moscow Aviation Institute, Russia
Donghyun Kim	Kennesaw State University, USA
Roman Kolpakov	Moscow State University, Russia
Alexander Kononov	Sobolev Institute of Mathematics, Russia
Igor Konnov	Kazan Federal University, Russia

Vera Kovacevic-Vujcic	University of Belgrade, Serbia
Yury A. Kochetov	Sobolev Institute of Mathematics, Russia
Pavlo A. Krokhmal	University of Arizona, USA
Ilya Kurochkin	Institute for Information Transmission Problems, RAS, Russia
Dmitri E. Kvasov	University of Calabria, Italy
Alexander A. Lazarev	V.A. Trapeznikov Institute of Control Sciences, Russia
Vadim Levit	Ariel University, Israel
Bertrand M. T. Lin	National Chiao Tung University, Taiwan
Alexander V. Lotov	Dorodnicyn Computing Centre, FRC CSC RAS, Russia
Olga Masina	Yelets State University, Russia
Vladimir Mazalov	Institute of Applied Mathematical Research, Karelian Research Center, Russia
Nevena Mijajlović	University of Montenegro, Montenegro
Mikhail Myagkov	University of Oregon, USA
Angelia Nedich	University of Illinois at Urbana Champaign, USA
Yuri Nesterov	CORE, Université Catholique de Louvain, Belgium
Yuri Nikulin	University of Turku, Finland
Evgeni Nurminski	Far Eastern Federal University, Russia
Panos Pardalos	University of Florida, USA
Alexander V. Pesterev	V.A. Trapeznikov Institute of Control Sciences, Russia
Alexander Petunin	Ural Federal University, Russia
Stefan Pickl	Uni Bw Munich, Germany
Boris T. Polyak	V.A. Trapeznikov Institute of Control Sciences, Russia
Leonid Popov	IMM UB RAS, Russia
Mikhail A. Posypkin	Dorodnicyn Computing Centre, FRC CSC RAS, Russia
Alexander N. Prokopenya	Warsaw University of Life Sciences, Poland
Oleg Prokopyev	University of Pittsburgh, USA
Artem Pyatkin	Novosibirsk State University; Sobolev Institute of Mathematics, Russia
Radu Ioan Boţ	University of Vienna, Austria
Soumyendu Raha	Indian Institute of Science, India
Leonidas Sakalauskas	Institute of Mathematics and Informatics, Lithuania
Eugene Semenkin	Siberian State Aerospace University, Russia
Yaroslav D. Sergeyev	University of Calabria, Italy
Natalia Shakhlevich	University of Leeds, UK

Alexander A. Shananin	Moscow Institute of Physics and Technology, Russia
Angelo Sifaleras	University of Macedonia, Greece
Mathias Staudigl	Maastricht University, The Netherlands
Petro Stetsyuk	V.M. Glushkov Institute of Cybernetics, Ukraine
Fedor Stonyakin	V. I. Vernadsky Crimean Federal University, Russia
Alexander Strekalovskiy	Institute for System Dynamics and Control Theory, SB RAS, Russia
Vitaly Strusevich	University of Greenwich, UK
Michel Thera	University of Limoges, France
Tatiana Tchemisova	University of Aveiro, Portugal
Anna Tatarczak	Maria Curie-Skłodowska University, Poland
Alexey A. Tretyakov	Dorodnicyn Computing Centre, FRC CSC RAS, Russia
Stan Uryasev	University of Florida, USA
Frank Werner	Otto von Guericke University Magdeburg, Germany
Adrian Will	National Technological University, Argentina
Anatoly A. Zhigljavsky	Cardiff University, UK
Julius Žilinskas	Vilnius University Sydney, Lithuania
Yakov Zinder	University of Technology, Australia
Tatiana V. Zolotova	Financial University under the Government of the Russian Federation, Russia
Vladimir I. Zubov	Dorodnicyn Computing Centre, FRC CSC RAS, Russia
Anna V. Zykina	Omsk State Technical University, Russia

Organizing Committee Chairs

Milojica Jaćimović	Montenegrin Academy of Sciences and Arts, Montenegro
Yuri G. Evtushenko	Dorodnicyn Computing Centre, FRC CSC RAS, Russia
Nicholas N. Olenev	Dorodnicyn Computing Centre, FRC CSC RAS, Russia

Organizing Committee

| Alla Albu | Dorodnicyn Computing Centre, FRC CSC RAS, Russia |
| Natalia Burova | Dorodnicyn Computing Centre, FRC CSC RAS, Russia |

Invited Talks

Computational Approaches for Solving Systems of Nonlinear Equations

Panos M. Pardalos

University of Florida, USA
http://www.ise.ufl.edu/pardalos/, https://nnov.hse.ru/en/latna/

Abstract. Finding one or more solutions to a system of nonlinear equations (SNE) is a computationally hard problem with many applications in sciences and engineering. First, we will briefly discuss classical approaches for addressing (SNE). Then, we will discuss the various ways that a SNE can be transformed into an optimization problem, and we will introduce techniques that can be utilized to search for solutions to the global optimization problem that arises when the most common reformulation is performed. In addition, we will present computational results using different heuristics.

Degenerate Equality Constrained Optimization Problems and P-Regularity Theory

Alexey Tret'yakov

Systems Research Institute, Polish Academy of Sciences, Warsaw, Poland
https://www.researchgate.net/profile/Alexey_Tretyakov

Abstract. We consider necessary optimality conditions for optimization problems with equality constraints given in the operator form as $F(x) = 0$, where F is an operator between Banach spaces. The paper addresses the case when the Lagrange multiplier λ_0 associated with the objective function might be equal to zero.

If the equality constraints are not regular at some point x^* in the sense that the Fréchet derivative of F at x^* is not onto, then the point $z^* = (x^*, \lambda_0^*, \lambda^*)$ is a degenerate solution of the classical Lagrange system of optimality conditions $\mathcal{L}(x, \lambda_0, \lambda) = 0$, where x^* is a solution of the optimization problem and (λ_0^*, λ^*) is a corresponding generalized Lagrange multiplier. We derive new conditions that guarantee that z^* is a locally unique solution of the Lagrange system. We also introduce a modified Lagrange system and prove that z^* is its regular locally unique solution.

The modified Lagrange system introduced in the paper can be used as a basis for constructing numerical methods for solving degenerate optimization problems. Our results are based on the construction of p–regularity and are illustrated by examples.

This is joint work with Yuri Evtushenko, Olga Brezhneva, and Vlasta Malkova

Contents

Mathematical Programming

Decomposition Method for Solving the Quadratic Programming Problem in the Aircraft Assembly Modeling

Stanislav Baklanov$^{(\boxtimes)}$ [ID], Maria Stefanova [ID], Sergey Lupuleac [ID],
Julia Shinder [ID], and Artem Eliseev [ID]

Peter the Great St.Petersburg Polytechnic University, St.Petersburg 195251, Russia
baklanov.stas@gmail.com

Abstract. Aircraft assembly modeling requires solving the contact problem which can be reduced to the Quadratic Programming (QP) problem with dense ill-conditioned Hessian. For dozens of thousands of variables solving the QP problem requires huge time and RAM space. This paper suggests decomposing the problem into several sub-QP problems according to the geometric regions to reduce both solving time and memory requirements. The important feature of the considered QP problem is the density of the Hessian matrix which means that decomposed regions have mutual interference not only along their common boundary but instead on their whole areas. Suggested decomposition method solves sub-QP problems iteratively until the solution of the original QP is found. Convergence is proved under certain conditions. The results for test models of fuselage section joint and simultaneous joint of the upper and lower wing panels are presented.

Keywords: Assembly modeling · Quadratic programming · Decomposition method · Dense matrix

1 Introduction

The aeronautical structures such as an airframe are made of many compliant large-scale parts which often have a complex shape. An intensive use of mathematical simulation of the assembly process is the way to speed up the production and increase the quality of final product. In the last decade, the modeling approach involving consideration of the contact interaction between the compliant parts has been developed and applied to variation simulation and assembly optimization. The detailed review of the papers related to the subject is presented in [10,16].

The simulation of assembly process with regard to compliance and contact interaction of parts involves the multiple solving of contact problem. By means of

The research was supported by Russian Science Foundation (project No. 22-19-00062, https://rscf.ru/en/project/22-19-00062/).

N. Olenev et al. (Eds.): OPTIMA 2022, LNCS 13781, pp. 3–17, 2022.
https://doi.org/10.1007/978-3-031-22543-7_1

variational formulation and using static condensation [4, 18, 22, 24], the contact problem can be reformulated in the form of a quadratic programming problem ˙[6, 9] that allows one to use powerful QP methods. This numerical approach outlined in [24] has been updated and applied to the aircraft assembly simulation in [12] and subsequent works by the same research group [13, 14, 18, 19, 27]. A similar approach is used in [8, 11, 25, 26] for simulating the assembly process in automotive and aerospace industry. This approach allows the efficient solving of the QP problems with up to 10,000 unknowns, which is sufficient for modeling the assembly processes of many aircraft structures. However, for simulation of the assembly of some aviation structures (e.g., joint of fuselage sections, simultaneous joint of the upper and lower wing panels), much more unknowns are required. In this case, it seems natural to use the approach related to domain decomposition. The present study is devoted to the development of such an approach as applied to assembly modeling problems.

When solving high-dimensional problems in mechanics, Domain Decomposition Methods (DDM) are widely used. For solving contact problems, non-overlapping methods based on the Finite Element Tearing and Interconnect (FETI) method have been developed by Vondrák et al. [23]. The application of DDM to a contact problem with friction is discussed in the paper of Oumaziz et al. [17]. The local assembly model with the use of one fastener to join bodies is studied by Blanzé et al. [3]. Roulet et al. have shown in [20] that DDM can help reduce computation time compared to standard commercial tools such as Abaqus, using the example of quasistatic problem with contact and friction. It should be noted that the DDM used to solve the contact problem is usually used in the following way. Each body in contact is considered as a subdomain or a group of subdomains, and the interaction between subdomains, as well as a possible contact area, are considered as interfaces. Thus, the problem in the subdomain becomes the problem of linear elasticity, and all non-linearities are localized at the interfaces. Contact problems in global assembly modeling have a number of significant differences: a dense reduced stiffness matrix (causing interfaces to be as large as subdomains themselves), large-scale bodies (the size of the corresponding subdomain of the body is still large in dimension), and the presence of disconnected contact regions (for example, in upper and lower wing-to-fuselage model). Along with the ability to use adaptive QP methods to solve contact problems quickly and efficiently in assembly problems, the idea of implementing DDM to solve related QP problems arises. The DDM proposed in the current paper for the QP problem is discussed in the further sections.

The paper is organized as follows. Section 2 describes the formulation of assembly problems as QP problems, Sect. 3 suggests decomposition method to solve this kind of problems. Its convergence and error estimations are discussed in Sect. 4. Aitken modification of the decomposition method is suggested in Sect. 5. Solving of sub-QP problems is discussed in Sect. 6. Numerical results for the decomposition method are presented in Sect. 7. The results are discussed in Sect. 8.

2 Problem Statement

The riveted and bolted joints are mainly used when assembling aircraft structures. Installed fasteners prevent large relative tangential displacements of the parts to be joined. It justifies the implementation of node-to-node contact model. Using the static condensation of the original stiffness matrix of the finite element model and the variational statement of the contact problem, the modeling of the assembly process can be reduced to solving the following problem [15]:

$$\min_{x} F(x) = \min_{x} \frac{1}{2} x^T K x - f^T x, \tag{1a}$$

$$\text{s. t. } Ax \leq g, \tag{1b}$$

where $x \in \mathbb{R}^n$ is the vector of displacements in the junction area (zone of possible contact) that are constrained by m nonpenetration conditions, $m \leq n$. The problem dimension n is much smaller than the dimension of original finite element model. Vector $f \in \mathbb{R}^n$ is the force vector. The reduced stiffness matrix $K \in \mathbb{R}^{n \times n}$ is symmetric, positive-definite and ill-conditioned. It has the block diagonal structure with dense blocks corresponding to assembled parts. Constraint matrix $A \in \mathbb{R}^{m \times n}$ is sparse. For assembly problems A defines the pairs of displacements related to each other by nonpenetration conditions, thus every row of the matrix contains 1 or 2 non-zero elements. Vector $g \in \mathbb{R}^m$ is the vector of initial gap between the parts. The feasible region has an inner point.

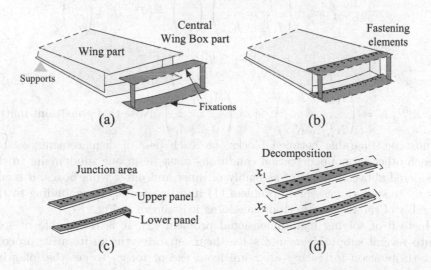

Fig. 1. Assembly of the outer and central wing boxes

The construction of the assembly model is illustrated on the example of the joint of the outer and central wing box, shown in Fig. 1(a) separately, and in Fig. 1(b) assembled. The procedure described above makes it possible to reduce the modeling of the assembly of these parts to solving problem (1), where only the junction area shown in Fig. 1(c) is considered.

The mechanical properties of the structure, and the configuration of the joining parts are characterized by the stiffness matrix K and the constraint matrix A. When performing a series of computations, these matrices do not change unlike vectors f and g. If the number of unknowns n is not very large (less than 10^4), the problem (1) is suitable for serial computations that take place in variation simulation, assembly optimization, etc. But as it was shown in [21], the time to solve problem (1) grows more than quadratically with the increase of n. Therefore, for problems with a large number of unknowns, the direct solution of problem (1) with QP methods ceases to be effective. At the same time, in some practically important problems that arise when modeling the assembly of aircraft structures, the number of unknowns cannot be made less than 10^4 while maintaining the required accuracy. In this regard, there is a need for development of a decomposition method, as applied for this class of problems.

3   Decomposition Method

Problem (1) is formulated in block form, corresponding to the subdomains of the original junction area:

$$\min_{x} \frac{1}{2}(x_1^T...x_r^T) \begin{pmatrix} K_{11} & ... & K_{1r} \\ ... & ... & ... \\ K_{1r}^T & ... & K_{rr} \end{pmatrix} \begin{pmatrix} x_1 \\ ... \\ x_r \end{pmatrix} - (f_1^T...f_r^T) \begin{pmatrix} x_1 \\ ... \\ x_r \end{pmatrix} \qquad (2a)$$

$$\text{s. t. } A_j x_j \le g_j, j = 1..r \qquad (2b)$$

where $x^T = (x_1^T...x_r^T)$, $x_j \in \mathbb{R}^{n_j}$, $f^T = (f_1^T...f_r^T)$, $f_j \in \mathbb{R}^{n_j}$, $g^T = (g_1^T...g_r^T)$, $g_j \in \mathbb{R}^{m_j}$, $K = \begin{pmatrix} K_{11} & ... & K_{1r} \\ ... & ... & ... \\ K_{1r}^T & ... & K_{rr} \end{pmatrix}$, $A = \begin{pmatrix} A_1 & ... & 0 \\ ... & ... & ... \\ 0 & ... & A_r \end{pmatrix}$. Note that constraint matrix A must be separable between blocks, i.e. each pair of displacements related to each other by nonpenetration conditions must lie in one subdomain. In the considered above example of assembly of inner and outer wing boxes, it seems natural to split the assembly problem (1) into two parts, corresponding to the assembly of the top and bottom panels, as it is shown in Fig. 1(d).

Instead of solving high-dimensional problem (2), it is reasonable to split it into several subproblems and solve them separately in an iterative process. At k-th iteration for each j all x are fixed except for x_j to get the following subproblem:

$$x_j^{k+1} = \arg \min_{x_j} F_j^k(x_j) = \arg \min_{x_j} \frac{1}{2} x_j^T K_{jj} x_j - f_j^{k^T} x_j \qquad (3a)$$

$$\text{s.t. } A_j x_j \le g_j, \qquad (3b)$$

where

$$f_j^k = f_j - \sum_{l=1, l\neq j}^{r} K_{jl}x_l^k \tag{4}$$

are corrective forces, defining mutual interference of subproblems with each other. It is worth noticing that unlike common DDM, matrices K_{jl} are not sparse. After solving each of the subproblems once, corrective forces will change, requiring to solve all subproblems again. Note that subproblems can be solved in parallel. Algorithm 1 describes the resulting iterative process. Stop criteria will be discussed in Sect. 4.

Algorithm 1. Decomposition method algorithm

1: $k \Leftarrow 0$, $x^0 \Leftarrow 0$
2: **while** stop criteria is not satisfied **do**
3:     **for** $j = 1$ to r **do**
4:         Set forces f_j^k using (4)
5:         Set x_j^{k+1} by solving subproblem (3)
6:     **end for**
7:     $k \Leftarrow k + 1$
8: **end while**

The approach similar to Gauss—Seidel method can be applied for acceleration of the convergence. At each iteration corrective forces can be updated using already calculated x_j^{k+1}:

$$f_j^k = f_j - \sum_{l=1}^{j-1} K_{jl}x_l^{k+1} - \sum_{l=j+1}^{r} K_{jl}x_l^k. \tag{5}$$

However, the multithreading is not possible with this modification.

4 Convergence

Denote $x^k = \begin{pmatrix} x_1^k \\ ... \\ x_r^k \end{pmatrix}$, $f^k = \begin{pmatrix} f_1^k \\ ... \\ f_r^k \end{pmatrix}$; $K_D = \begin{pmatrix} K_{11} ... & 0 \\ ... & ... & ... \\ 0 & ... K_{rr} \end{pmatrix}$ is block diagonal part of matrix K; $K_S = K - K_D$ is side part of matrix K. Thus, $f^k = f - K_S x^k$.

Theorem 1. *Algorithm 1 converges to the minimum of problem (1) linearly with the convergence rate*

$$q = 2\frac{|\lambda|_{max}(K_S)}{\lambda_{min}(K_D)} \tag{6}$$

if $q < 1$, meaning the predominance of the block diagonal part K_D over non-diagonal part K_S.

8   S. Baklanov et al.

Proof. First, the function difference $\Delta^k F_j^k = F_j^k(x_j^{k+1}) - F_j^k(x_j^k)$ is estimated from below for any j. By definition of the minimum for $(k-1)^{th}$ iteration:

$$F_j^{k-1}(x_j^{k+1}) \geq F_j^{k-1}(x_j^k) \tag{7}$$

Adding $F_j^k(x_j^{k+1}) - F_j^{k-1}(x_j^{k+1}) = f_j^{k-1^T} x_j^{k+1} - f_j^{k^T} x_j^{k+1}$ to both parts:

$$F_j^k(x_j^{k+1}) \geq \frac{1}{2} x_j^{k^T} K_{jj} x_j^k - f_j^{k-1^T} x_j^k + f_j^{k-1^T} x_j^{k+1} - f_j^{k^T} x_j^{k+1}. \tag{8}$$

Subtracting $F_j^k(x_j^k)$ from both parts:

$$\Delta^k F_j^k \geq -f_j^{k-1^T} x_j^k + f_j^{k-1^T} x_j^{k+1} - f_j^{k^T} x_j^{k+1} + f_j^{k^T} x_j^k = -\Delta^{k-1} f_j^T \Delta^k x_j, \tag{9}$$

where $\Delta^{k-1} f_j = f_j^k - f_j^{k-1}$, $\Delta^k x_j = x_j^{k+1} - x_j^k$.

Second, the function difference is estimated from above using $\Delta^k x_j$:

$$\Delta^k F_j^k = -\frac{1}{2} \Delta^k x_j^T K_{jj} \Delta^k x_j - (x_j^{k+1^T} K_{jj}(-\Delta^k x_j) - f_j^{k^T}(-\Delta^k x_j)). \tag{10}$$

Segment between x_j^k and x_j^{k+1} lies in the feasible region (as both x_j^k and x_j^{k+1} satisfy linear constraints), hence directional derivative of F_j^k in its minimum point x_j^{k+1} in direction $-\Delta^k x_j$ must be non-negative:

$$\nabla^T F_j^k(x_j^{k+1})(-\Delta^k x_j) = x_j^{k+1^T} K_{jj}(-\Delta^k x_j) - f_j^{k^T}(-\Delta^k x_j) \geq 0. \tag{11}$$

Using (11) in (10):

$$\Delta^k F_j^k \leq -\frac{1}{2} \Delta^k x_j^T K_{jj} \Delta^k x_j. \tag{12}$$

Uniting (12) and (9):

$$\frac{1}{2} \Delta^k x_j^T K_{jj} \Delta^k x_j \leq \Delta^{k-1} f_j^T \Delta^k x_j. \tag{13}$$

$$\frac{1}{2} \lambda_{min}(K_{jj}) \|\Delta^k x_j\|_2^2 \leq \|\Delta^{k-1} f_j\|_2 \|\Delta^k x_j\|_2. \tag{14}$$

Thus, difference of x_j can be estimated using difference of forces f_j:

$$\|\Delta^k x_j\|_2 \leq \frac{2\|\Delta^{k-1} f_j\|_2}{\lambda_{min}(K_{jj})} \leq \frac{2\|\Delta^{k-1} f_j\|_2}{\lambda_{min}(K_D)}. \tag{15}$$

Uniting for all x_j:

$$\|\Delta^k x\|_2 \leq \frac{2\|\Delta^{k-1} f\|_2}{\lambda_{min}(K_D)} \leq \frac{2\|(f - K_S x^k) - (f - K_S x^{k-1})\|_2}{\lambda_{min}(K_D)} \leq$$
$$\frac{2|\lambda|_{max}(K_S)}{\lambda_{min}(K_D)} \|\Delta^{k-1} x\|_2. \tag{16}$$

Finally, iteration process converges linearly if

$$q = 2\frac{|\lambda|_{max}(K_S)}{\lambda_{min}(K_D)} < 1. \tag{17}$$

Assume that sequence $\{x^k\}$ given by Algorithm 1 converges to point x^*. The fact that this point is a solution to problem (1) is proved next.

The Karush—Kuhn—Tucker (KKT) necessary conditions for subproblem (3) are as follows:

$$\nabla F_j(x^*) + \lambda_j^T A_j = 0; A_j x_j^* \leq g_j; \lambda_j \geq 0; \lambda_j^T (A_j x_j^* - g_j) = 0, \tag{18}$$

where $\lambda_j \in \mathbb{R}^{m_j}$ are Lagrange multipliers representing contact forces, $\lambda^T = (\lambda_1^T ... \lambda_r^T)$. As subproblem (3) is a QP problem with positive definite Hessian matrix and, according to the problem statement, there exists inner point in the feasible region, KKT conditions are also sufficient.

Uniting KKT conditions for all j:

$$\nabla F(x^*) + \lambda^T A = 0; Ax^* \leq g; \lambda \geq 0; \lambda^T (Ax^* - g) = 0. \tag{19}$$

Note that these are KKT conditions for the main problem (1). Analogously, they are necessary and sufficient. Thus, x^* is a minimum point for (1).   □

As for any linear convergent sequence, accuracy of the solution can be estimated as

$$\|x^k - x^*\|_2 \leq \frac{q}{1-q}\|x^k - x^{k-1}\|_2 < \epsilon. \tag{20}$$

Stop criteria for Algorithm 1 can be set according to (20), where

$$q \approx \frac{\|x^k - x^{k-1}\|_2}{\|x^{k-1} - x^{k-2}\|_2} \tag{21}$$

Other approaches for stop criteria are using the number of iterations, difference $\|x^k - x^{k-1}\|_2$, or KKT conditions.

Moreover, accuracy of the solution can be estimated directly. Vector x_j^{k+1} is a minimum point for QP subproblem (3) if and only if all directional derivatives in all feasible directions are non-negative:

$$\nabla^T F_j^k(x_j^{k+1})d_j \geq 0, \tag{22}$$

where d_j is a sufficiently small direction vector of length n_j, and $x_j^{k+1} + d_j$ satisfies subproblem constraints (3b).

$$(K_{jj}x_j^{k+1} - f_j^k)^T d_j \geq 0. \tag{23}$$

Uniting these equations for all j:

$$(K_D x^{k+1} - f^k)^T d \geq 0. \tag{24}$$

$$(Kx^{k+1} - f + K_S(x^k - x^{k+1}))^T d \geq 0. \tag{25}$$

$$\nabla^T F(x^{k+1})d \geq d^T K_S(x^{k+1} - x^k). \tag{26}$$

Let $d = x^* - x^{k+1}$. Then

$$0 \leq F(x^{k+1}) - F(x^*) = -\frac{1}{2}d^T Kd - x^{k+1}{}^T Kd + f^T d. \tag{27}$$

$$\frac{1}{2}d^T Kd \leq -\nabla^T F(x^{k+1})d \leq -d^T K_S(x^{k+1} - x^k) = d^T(f^{k+1} - f^k). \tag{28}$$

$$\frac{1}{2}\lambda_{min}(K)\|d\|_2^2 \leq \|d\|_2\|f^{k+1} - f^k\|_2. \tag{29}$$

$$\|x^* - x^{k+1}\|_2 \leq \frac{2}{\lambda_{min}(K)}\|f^{k+1} - f^k\|_2. \tag{30}$$

Thus, accuracy of the solution can be estimated via difference of forces at consequent iterations.

5   Aitken Modification

As it was shown earlier, Algorithm 1 converges linearly under certain conditions. One may consider the repetitive part of the algorithm as a fixed-point iteration [7] for improvement of the convergence rate. Two operators are introduced for Eqs. (4) and (3), respectively:

$$\begin{cases} f^k = C(x^k) \\ x^{k+1} = D(f^k) \end{cases} \tag{31}$$

Note that C is a linear operator and D provides a vector filled with minimizers of the corresponding QP subproblems.

The force vector f^k may be excluded from (31), providing the fixed-point formulation of Algorithm 1:

$$x^{k+1} = D(C(x^k)) = S(x^k), \tag{32}$$

where operator S represents one step of the decomposition method.

The numerical performance of the iterative procedure (32) may be improved, if it is replaced by relaxation procedure:

$$x^{k+1} = x^k + \omega^k(y^k - x^k), \tag{33}$$

where $y^k = S(x^k)$ and ω^k is a so-called relaxation parameter, which, in general, varies between the iterations.

The improvement is achieved, when appropriate formulation for the relaxation parameter ω^k is specified. For that purpose, one may use the generalization of the secant method introduced in [5]:

$$\omega^k = -\omega^{k-1}\frac{(y^{k-1} - x^{k-1})^T((y^k - x^k) - (y^{k-1} - x^{k-1}))}{\|(y^k - x^k) - (y^{k-1} - x^{k-1})\|_2^2}, \tag{34}$$

where $\omega^0 \in (0,1]$, e.g. $\omega^0 = 1$. Relaxation procedure (33), (34) is usually called Aitken relaxation method [7], since it was introduced originally as a modification of the Aitken delta-squared procedure.

It is worth noting that Aitken relaxation should be used with caution as x^{k+1} achieved with (33) may not satisfy constraints (1b). Nonetheless, displacements x^{k+1} can still be used for calculating corrective forces with (4). However, at the last iteration either x^{k+1} should be projected on the feasible set or y^k should be used as an answer.

Finally, decomposition method with Aitken modification is described in Algorithm 2 and its influence on the convergence is described in Sect. 7. Unfortunately, the error estimation (20) is no longer applicable, however, (30) still stands for y^k and can be modified for the new x^{k+1}:

$$\|x^* - x^{k+1}\|_2 \leq \|x^* - y^k\|_2 + \|x^{k+1} - y^k\|_2 \leq$$

$$\frac{2}{\lambda_{min}(K)}\|f^{k+1} - f^k\|_2 + |1 - \omega^k| \cdot \|(x^k - y^k)\|_2. \quad (35)$$

Algorithm 2. Decomposition method algorithm with Aitken modification

---

1: $k \Leftarrow 0$, $x^0 \Leftarrow 0$, $\omega^0 \Leftarrow 1$
2: **while** Stop criteria is not satisfied **do**
3:     **for** $j = 1$ to r **do**
4:         Set forces f_j^k using (4)
5:         Set y_j^k by solving subproblem (3)
6:     **end for**
7:     Set ω^k using (34)
8:     Set x^{k+1} using (33)
9:     $k \Leftarrow k + 1$
10: **end while**

---

6 Solution Methods of Quadratic Programming Subproblems

Any method for solving QP problem can be used for solving subproblems (3). It is even possible to use different methods for different subproblems or change methods at consequent iterations. The following methods are considered [21]: Goldfarb—Idnani active set method, Newton projection method, primal-dual interior point method and Lemke's method. The first two show the best performance in the assembly modeling. Apart from solving subproblems (3) (called primal formulation of QP) directly, their dual or so-called relative equivalent formulations can be used as they can have reduced solving time [1,21].

Solving subproblems (3) is the most time-consuming part of the decomposition method, where time can be roughly estimated as cubic with respect to n_j [21]. For each subproblem solving it from the beginning for every iteration may be inefficient. Note that subproblems at consequent iterations differ only by the force vector f_j^k, and the bigger the number of iteration k is, the smaller is the difference of forces $\|f_j^{k+1} - f_j^k\|_2$ and the difference of solutions $\|x_j^{k+1} - x_j^k\|_2$ between iterations. Thus, using previous solution x_j^{k-1} (or y_j^{k-1} for Aitken modification) at k iteration as a warm start (i.e. initial estimate) may reduce the solving time significantly. Indeed, for x_j^k sufficiently close to x_j^* the correct set of active constraints will be determined most surely and subproblem (3) will be effectively reduced to unconstrained minimization with fixed Hessian [1,2], thus reducing solving time to quadratic with respect to n_j. The effectiveness of warm-start usage is shown in Sect. 7.

7   Results

Decomposition method (Algorithm 2) is implemented in ASRP software complex [21,28]. Results for two best performing methods are shown: Goldfarb—Idnani active set method (ASM) and Newton projection method (NPM). Letters "P", "D" stand for primal and dual formulations [1,21], respectively.

Decomposition method is tested on four benchmark models. The first two benchmarks B1 and B2 (Fig. 2) are based on the outer and center wing box assembly model, shown schematically in Fig. 1. The benchmarks differ only in the density of the computational grid: B1 contains 6652, and B2 contains 14580 unknowns, respectively. The decomposition for these benchmarks consists of dividing the junction area into joints of the upper and lower panels, as it is shown in Fig. 1(d).

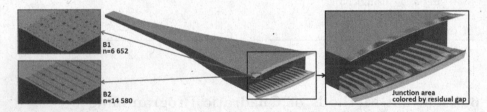

Fig. 2. Benchmarks B1 and B2 based on the outer and center wing box assembly model

The two other benchmarks B3 and B4 are built on the base of model of assembly of half fuselage sections (Fig. 3). They also differ only in the computational grid density: B3 contains 8316, and B4 contains 31450 unknowns, respectively. Decomposition is carried out as shown in Fig. 3 on the right.

Solving times for benchmarks B1 and B2 are illustrated in Fig. 4 with the dependency on the number of active constraints at optimum. Time is shown in logarithmic scale. For all QP solvers decomposition method converges in 3 or 4 iterations and shows significant improvement of the solving time compared to solving problem (1) directly. Fast convergence is the result of the small dependency between two subdomains, i.e. significant predominance of diagonal matrix K_D over nondiagonal K_S (Theorem 1).

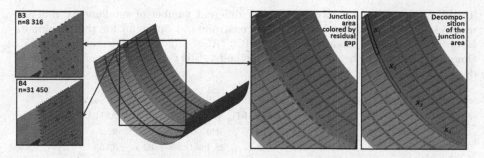

Fig. 3. Benchmarks B3 and B4 based on the model of assembly of half fuselage sections

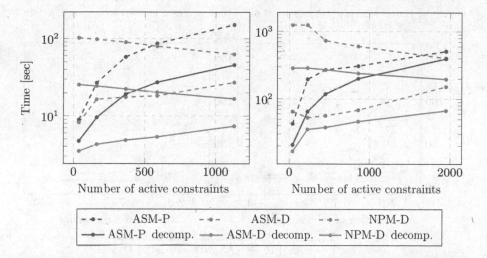

Fig. 4. Comparison of solving time for different QP solving methods without and with decomposition method depending on number of active constraints at optimum for wing models B1 (left) and B2 (right). 'decomp.' stands for decomposition method

For the benchmark B3 Table 1 and Fig. 5 compare solving time for different methods with different number of subdomains. Unlike previous model, subdomains represent adjoining zones of the junction area (see Fig. 3, right). For $r = 1$ problem (1) is solved without decomposition method. Usage of the decomposition method still significantly reduces solving time, and the more subdomains are used, the smaller is the solving time until some critical value for subdomains is reached. It is also worth noticing that the number of decomposition method iterations usually decreases as the number of active constraints at optimum increases.

Table 2 shows difference in solving time for decomposition method with and without warm start usage for the benchmark B3 with 7 subdomains. Warm start helps to decrease time by about ten times.

Unfortunately, for the benchmark B4 with 7 subdomains (fastest case for B3) convergence is slow and decomposition method may require hundreds or thou-

Table 1. Comparison of solving times for different number of subdomains, different methods and different number of active constraints at optimum for the benchmark B3. r is a number of subdomains. "ASM-P", "ASM-D", "NPM-D" stand for solving times in seconds for the corresponding methods. "Iter." is a number of iterations of decomposition method iterations

r	196 active constraints				1130 active constraints				1554 active constraints			
---	ASM-P	ASM-D	NPM-D	Iter	ASM-P	ASM-D	NPM-D	Iter	ASM-P	ASM-D	NPM-D	Iter
1	33.4	166	10.5	-	216	109	51.9	-	274	92.2	94.4	-
2	55.8	51.7	7.8	50	106	42.1	28.4	53	102	32.5	33.6	30
3	44.8	31.1	8.6	69	52.2	19.6	18.6	39	60.5	19	23.6	39
6	19.6	11.2	6.6	64	22.1	9.2	7.4	51	20.2	7.6	6.6	39
7	17.3	10.5	6.9	70	16.2	7.2	6.1	44	17.5	7.4	6.7	47
14	14.5	11.6	12.1	117	9.8	7.6	6.8	68	11.5	8.7	8	79

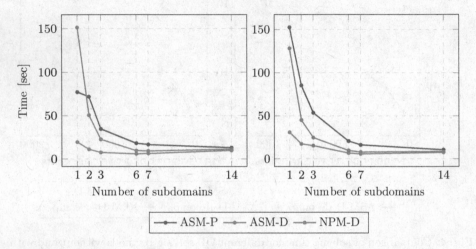

Fig. 5. Comparison of solving time for different QP solving methods depending on number of subdomains for the coarse fuselage model B3. Number of active constraints at optimum is 376 (left) and 736 (right)

Table 2. Comparison of solving time for the decomposition method w/o and with warm start usage for benchmark B3 with 7 subdomains

	ASM-P					NPM-P				
Active constraints	196	376	736	1130	1543	196	376	736	1130	1543
Time w/o warm start, s	140	298	294	729	729	1228	891	539	575	624
Time with warm start, s	35	21	23	35	27	42	39	26	27	28

sands of iterations or even do not converge at all. This result is the consequence of the Theorem 1 as well, as the mutual influence of subdomains is significant. Possible solutions to this problem will be discussed in Sect. 8. Still, some analysis can be made on this model.

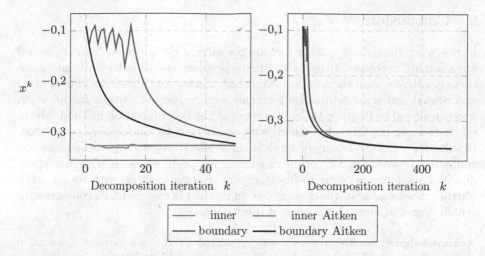

Fig. 6. Comparison of convergence for components of x^k corresponding to inner and boundary nodes of subdomains without and with Aitken modification for the benchmark B4. Left graphic shows first 50 iterations, and right graphic shows first 500 iterations

Figure 6 shows convergence of components of x^k lying in the inner or boundary part of the subdomains with and without Aitken modification usage. Left graphic shows first 50 iterations, and right graphic shows first 500 iterations. One can see that inner components converge quickly and decent result is achieved in the first iterations of the decomposition method. However, boundary parts converge very slowly, requiring hundreds of iterations to converge because of the interference from a neighbor part. Aitken modification helps to avoid oscillations in the first iterations of the decomposition method, thus improving convergence locally.

8   Discussion

Future work is possible on improving convergence rate of the decomposition method from linear to quadratic. As mentioned in Sect. 6, the correct set of active constraints is determined starting from some iteration. Then $\Delta^k x$ is actually linearly dependent on $\Delta^{k-1} f$ and, thus on $\Delta^{k-1} x$. This may allow usage of second order method upon $\{x^k\}$ to achieve x^* faster.

Another possible approach to improve the convergence is to divide the junction area in two ways with different splitting patterns in order to utilize the fact that components of displacement vector x converge faster if those components lie in the inner part of the subdomain (see Fig. 6). Then even if some component lies on the boundary for one set of subdomains, it would lie in the inner part for another set of subdomains. By alternation of two subsets of subdomains and corresponding subproblems during iterations of Algorithm 1, faster convergence may be expected.

9   Conclusion

In this work the decomposition method for solving QP problems was suggested for assembly problems. It splits the main problem into smaller ones and solve them iteratively until the solution is reached. Linear convergence of this method was proved and error estimation formulae were suggested. Aitken modification was considered to improve the convergence of the decomposition method. Methods for solving the QP subproblems with warm-start technique were considered. Results for different assembly models show effectiveness of proposed method compared to ordinary QP problem solving for problems with a small dependence of subdomains (wing to fuselage joint, fuselage section with rough grid). Further development of the decomposition method is required for problems with a high degree of mutual influence of subdomains.

Acknowledgments. The research was supported by Russian Science Foundation (project No. 22-19-00062, https://rscf.ru/en/project/22-19-00062/).

References

1. Baklanov, S., Stefanova, M., Lupuleac, S.: Newton projection method as applied to assembly simulation. Optim. Methods Softw. **37**, 1–28 (2020)
2. Bertsekas, D.: Projected newton methods for optimization problems with simple constraints. J. Control Optim. **20**(2), 221–246 (1982)
3. Blanzé, C., Champaney, L., Cognard, J.Y., Ladeveze, P.: A modular approach to structure assembly computations: application to contact problems. Eng. Comput. **13**, 15–32 (1996)
4. Guyan, R.J.: Reduction of stiffness and mass matrix. AIAA J. **3**(2), 380 (1965)
5. Irons, B., Tuck, R.: A version of the Aitken accelerator for computer implementation. Int. J. Numer. Methods Eng. **1**, 275–277 (1969)
6. Kinderlehrer, D., Stampacchia, G.: An Introduction To Variational Inequalities and their Applications. SIAM, Philadelphia (1980)
7. Küttler, U., Wall, W.A.: Fixed-point fluid-structure interaction solvers with dynamic relaxation. Comput. Mech. **42**, 61–72 (2008)
8. Lindau, B., Lorin, S., Lindkvist, L., Söderberg, R.: Efficient contact modeling in nonrigid variation simulation. ASME J. Comput. Inf. Sci. Eng. **16**(1), 1–7 (2016)
9. Lions, J.L., Stampacchia, G.: Variational inequalities. Commun. Pure Appl. Math. **20**, 493–519 (1967)
10. Lorin, S., Lindau, B., Lindkvist, L., Söderberg, R.: Efficient compliant variation simulation of spot-welded assemblies. ASME. J. Comput. Inf. Sci. Eng. **19**(1), 011007 (2019)
11. Lorin, S., Lindau, B., Tabar, R.S., Lindkvist, L., Wärmefjord, K., Söderberg, R.: Efficient variation simulation of spot-welded assemblies. In: ASME International Mechanical Engineering Congress and Exposition, Proceedings (IMECE), vol. 2, ASME (2018)
12. Lupuleac, S., Kovtun, M., Rodionova, O., Marguet, B.: Assembly simulation of riveting process. SAE Int. J. Aerosp. **2**, 193–198 (2010)
13. Lupuleac, S., Petukhova, M., Shinder, Y., Bretagnol, B.: Methodology for solving contact problem during riveting process. SAE Int. J. Aerosp. **4**(2), 952–957 (2011)

14. Lupuleac, S., Shinder, Y., Petukhova, M., Yakunin, S., Smirnov, A., Bondarenko, D.: Development of numerical methods for simulation of airframe assembly process. SAE Int. J. Aerosp. **6**(1), 101–105 (2013)
15. Lupuleac, S., Smirnov, A., Churilova, M., Shinder, J., Zaitseva, N., Bonhomme, E.: Simulation of body force impact on the assembly process of aircraft parts. In: ASME International Mechanical Engineering Congress and Exposition. American Society of Mechanical Engineers (2019)
16. Lupuleac, S., et al.: Simulation of the wing-to-fuselage assembly process. ASME. J. Manuf. Sci. Eng. **141**(6), 061009 (2019)
17. Oumaziz, P., Gosselet, P., Boucard, P.A., Abbas, M.: A parallel non-invasive multi-scale strategy for the mixed domain decomposition method with frictional contact. Int. J. Numer. Methods Eng. **115**, 1–14 (2018)
18. Petukhova, M., Lupuleac, S., Shinder, Y., Smirnov, A., Yakunin, S., Bretagnol, B.: Numerical approach for airframe assembly simulation. J. Math. Ind. **4**(8), 1–2 (2014)
19. Pogarskaia, T., Churilova, M., Petukhova, M., Petukhov, E.: Simulation and optimization of aircraft assembly process using supercomputer technologies. In: Voevodin, V., Sobolev, S. (eds.) RuSCDays 2018. CCIS, vol. 965, pp. 367–378. Springer, Cham (2019). https://doi.org/10.1007/978-3-030-05807-4_31
20. Roulet, V., Champaney, L., Boucard, P.A.: A parallel strategy for the multiparametric analysis of structures with large contact and friction surfaces. Adv. Eng. Softw. **42**(6), 347–368 (2011)
21. Stefanova, M., et al.: Convex optimization techniques in compliant assembly simulation. Optim. Eng. **21**, 1–26 (2020). https://doi.org/10.1007/s11081-020-09493-z
22. Turner, M.J., Clough, R.W.C., M.H., Topp, L.J.: Stiffness and deflection analysis of complex structures. J. Aero. Sci. **23**, 805–823 (1956)
23. Vondrák, V., Dostal, Z., Dobiash, J., Ptak, S.: A FETI domain decomposition method applied to contact problems with large displacements. In: Widlund, O.B., Keyes, D.E. (eds.) Domain Decomposition Methods in Science and Engineering XVI. LNCSE, vol. 55. Springer, Berlin, Heidelberg (2007). https://doi.org/10.1007/978-3-540-34469-8_96
24. Wriggers, C.P.: Computational Contact Mechanics. Springer Verlag, Heidelberg (2006). https://doi.org/10.1007/978-3-540-32609-0
25. Wärmefjord, K., Söderberg, R., Lindau, B., Lindkvist, L., Lorin, S.: Joining in nonrigid variation simulation. In: Udroiu, R. (ed.) Computer-aided Technologies IntechOpen, London (2016)
26. Yang, D., Qu, W., Ke, Y.: Evaluation of residual clearance after pre-joining and pre-joining scheme optimization in aircraft panel assembly. Assem. Autom. **36**(4), 376–387 (2016)
27. Zaitseva, N., Lupuleac, S., Petukhova, M., Churilova, M., Pogarskaia, T., Stefanova, M.: High performance computing for aircraft assembly optimization. In: 2018 Global Smart Industry Conference (GloSIC), pp. 1–6. IEEE (2018)
28. Zaitseva, N., Pogarskaia, T., Minevich, O., Shinder, J., Bonhomme, E.: Simulation of aircraft assembly via ASRP software. Technical report, SAE Technical Paper (2019)

Degenerate Equality Constrained Optimization Problems and P-Regularity Theory

Olga Brezhneva[1], Yuri Evtushenko[2,3] (iD), Vlasta Malkova[2(✉)] (iD),
and Alexey Tret'yakov[2,4,5] (iD)

[1] Department of Mathematics, Miami University, Oxford, OH 45056, USA
brezhnoa@miamioh.edu
[2] FRC CSC RAS, Vavilov Street 40, 119333 Moscow, Russia
vmalkova@yandex.ru
[3] Moscow Institute of Physics and Technology, Moscow, Russia
yuri-evtushenko@yandex.ru
[4] System Research Institute, Polish Academy of Sciences, Newelska 6,
01-447 Warsaw, Poland
[5] Faculty of Exact and Natural Sciences, Siedlce University, 08-110 Siedlce, Poland
tret@ap.siedlce.pl

Abstract. We consider necessary optimality conditions for optimization problems with equality constraints given in the operator form as $F(x) = 0$, where F is an operator between Banach spaces. The paper addresses the case when the Lagrange multiplier λ_0 associated with the objective function might be equal to zero. If the equality constraints are not regular at some point x^* in the sense that the Fréchet derivative of F at x^* is not onto, then the point $z^* = (x^*, \lambda_0^*, \lambda^*)$ is a degenerate solution of the classical Lagrange system of optimality conditions $\mathcal{L}(x, \lambda_0, \lambda) = 0$, where x^* is a solution of the optimization problem and (λ_0^*, λ^*) is a corresponding generalized Lagrange multiplier. We derive new conditions that guarantee that z^* is a locally unique solution of the Lagrange system. We also introduce a modified Lagrange system and prove that z^* is its regular locally unique solution. The modified Lagrange system introduced in the paper can be used as a basis for constructing numerical methods for solving degenerate optimization problems. Our results are based on the construction of p–regularity and are illustrated by examples.

Keywords: Equality constrained optimization problems ·
Degeneracy · Generalized necessary conditions

1 Introduction

We consider the nonlinear optimization problem with equality constraints:

$$\text{minimize } f(x) \quad \text{subject to} \quad F(x) = 0, \tag{1}$$

This work was supported in part by the Russian Foundation for Basic Research, project No. 21-71-30005.

where a functional $f : X \to \mathbb{R}$ and an operator $F : X \to Y$ between Banach spaces X and Y are sufficiently smooth in some neighborhood of a local solution x^*.

The paper describes an application of the p-regularity theory to nonlinear programming problems (1) with mapping F singular at the solution point x^*.

Recently, nonlinear problems and among them optimization problems attract the attention of the specialists in various disciplines. Moreover, nonlinear problems are closely connected with singular problems, as it was shown in [18]. Indeed, it turned out that so-called essentially nonlinear problems and singular problems are locally equivalent. We try to substantiate it. For this purpose we introduce some auxiliary notions.

Definition 1. *Let V be a neighborhood of x^* in X and $U \subset X$ be a neighborhood of 0. A mapping $F \colon V \to Y$, $F \in C^2(V)$, is essentially nonlinear at x^* if there exists a perturbation of the form $\tilde{F}(x^* + x) = F(x^* + x) + \omega(x)$, where $\|\omega(x)\| = o(\|x\|)$, such that there does not exist any nondegenerate transformation of coordinates $\varphi(x)\colon U \to V$, $\varphi \in C^1(U)$ such that $\varphi(0) = x^*$, $\varphi'(0) = I_X$ and $\tilde{F}(\varphi(x)) = \tilde{F}(x^*) + \tilde{F}'(x^*)x$ for all $x \in U$ holds with φ and $\tilde{F}$.*

Definition 2. *We say the mapping F is singular (or degenerate, nonregular) at x^* if it fails to be regular, that is, its derivative is not onto:*

$$\operatorname{Im} F'(x^*) \neq Y.$$

The following Theorem 1, which establishes the relationship between these two notions of the essential nonlinearity and singularity, was derived in [18].

Theorem 1. *Suppose $F \colon V \to Y$ is C^2 and that x^* is a solution of $F(x) = 0$. Then F is essentially nonlinear at the point x^* if and only if F is singular at the point x^*.*

Therefore, in this paper we want to draw attention to what is the main feature of the nonlinearity problems by means of nonlinear programming problems with a singularity in the solution.

The Lagrangian for problem (1) is defined as

$$L(x, \lambda_0, \lambda) = \lambda_0 f(x) + \langle \lambda, F(x) \rangle,$$

where $(\lambda_0, \lambda) \in (\mathbb{R} \times Y^*) \backslash \{0\}$ is a generalized Lagrange multiplier.

The classical first-order Euler-Lagrange necessary optimality conditions for problem (1) state that if x^* is a local minimizer of (1) and the image of $F'(x^*)$ is closed, then there exists a generalized Lagrange multiplier (λ_0^*, λ^*) such that [1]:

$$\lambda_0^* f'(x^*) + (F'(x^*))^* \lambda^* = 0, \tag{2}$$

where not all of λ_0^*, λ^* are zero. In other words, the point $(x^*, \lambda_0^*, \lambda^*)$ is a solution of the following system of equations:

$$\mathcal{L}(x, \lambda_0, \lambda) = \begin{pmatrix} \lambda_0 f'(x) + (F'(x))^* \lambda \\ F(x) \\ \lambda_0^2 + \|\lambda\|^2 - 1 \end{pmatrix} = 0, \tag{3}$$

where $\mathcal{L}$ is a mapping from $X \times \mathbb{R} \times Y^*$ to $X^* \times Y \times \mathbb{R}$. We assume that the derivative of $\|\lambda\|^2$ exists and is not zero if $\lambda \neq 0$. System (3) is called the *Lagrange system*.

We are interested in the nonregular case, when the Lagrange multiplier λ_0^* might be equal to zero and the operator $F'(x^*)$ is not onto. In this case, the mapping $\mathcal{L}$ and system (3) are degenerate at $(x^*, 0, \lambda^*)$. Of course, necessary conditions (2) can be satisfied in a trivial way for any nonregular point x^*. Namely, assuming that $\operatorname{Im} F'(x^*)$ is closed, it is possible simply to choose $\lambda^* \neq 0$ such that $(F'(x^*))^* \lambda^* = 0$, and this choice will satisfy (2) with $\lambda_0^* = 0$. One of the goals of this paper is to construct a modification of system (3) that has the same point $(x^*, 0, \lambda^*)$ as its regular solution.

Motivation for our consideration is the problems for which the optimality conditions (2) are not satisfied with $\lambda_0^* \neq 0$. One such example is

$$\underset{(x_1, x_2)}{\text{minimize}} \ x_2 \quad \text{subject to} \quad F(x_1, x_2) = x_1^2 - x_2^3 = 0.$$

We consider this problem in detail in Example 1 in Sect. 4 of the paper. We are also *motivated* by the problems for which the point $z^* = (x^*, \lambda_0^*, \lambda^*)$ is a singular (degenerate) solution of Lagrange system (3). Namely, the numerical methods, which are based on the system of optimality conditions, either do not converge or loose their rate of convergence for degenerate problems. The modified Lagrange system introduced in the paper might be used as a basis for developing numerical methods for solving degenerate optimization problems.

The development of optimality conditions for nonregular problems has become an active research topic (see [2–14] and references therein). Here we pursue an approach based on the construction of p-regularity introduced earlier in [15–18]. In previous papers [2,7,15,16], optimality conditions for nonregular optimization problems with equality constraints were obtained for the case when the mapping F is p-regular at x^*. The main idea in [2,7,15,16] was to replace the operator $F'(x^*)$ which is not onto with a linear operator $\Psi_p(x^*)$, related to the pth order Taylor polynomial of F at x^*, which is onto. The operator $\Psi_p(x^*)$ contains up to the p-th order derivative of F, so F is p-times continuously Fréchet differentiable in a neighborhood of x^*. The order p was chosen as the minimum number for which the operator $\Psi_p(x^*)$ is regular.

The *novelty* of the paper is in analysis of properties of the mapping $\mathcal{L}$ defined in (3) by using the p-regularity theory [18]. We used a similar idea in [19], where a modified Lagrange system was introduced for inequality constrained optimization problems in the finite dimensional spaces to construct a new method with a superlinear rate of convergence to a degenerate solution. In this paper, we use a modified Lagrange system to derive new optimality conditions for nonregular optimization problems with equality constraints.

The *main contribution* of the paper is in new generalized necessary conditions that are stated and proved as Theorems 4–7 in Sect. 3. To be more specific, Theorem 4 introduces conditions under which z^* is a regular solution of the modified Lagrange system $\bar{\mathcal{L}}(z) = 0$ introduced in (17). In addition, Theorem 4 states assumptions under which the solution z^* is locally unique. Theorem 5 provides generalized necessary optimality conditions for problem (1). Theorem 6 gives a specific form of the modified Lagrange system in the case of $p = 2$. While Theorems 4–6 are given in terms of the modified Lagrange system, Theorem 7 relates to the classical Lagrange system (3). Namely, Theorem 7 states new conditions under which z^* is a locally unique solution of (3).

Note that when equality constraints are not regular at the solution of the optimization problem, the classical optimality conditions are satisfied in a trivial way with $\lambda_0^* = 0$, so the term containing the objective function is eliminated from system (3). In contrast, the first- or a higher-order derivative of the objective function might be present explicitly with a nonzero multiplier in the modified optimality system. Moreover, results similar to ones presented in the paper for the case of $p = 2$ can be derived for the problems where F is a differentiable mapping whose derivative is Lipschitz continuous. This direction would also extend the results presented in [9].

The organization of the paper is as follows. In Sect. 2, we recall some definitions and results of the p-regularity theory [10,18]. In Sect. 3, we derive generalized necessary conditions. In Sect. 4, we illustrate the results by some examples.

Notation. Let $\mathcal{L}(X, Y)$ be the space of all continuous linear operators from X to Y and for a given linear operator $\Lambda : X \to Y$, we denote its kernel and image by $\mathrm{Ker}\Lambda = \{x \in X \mid \Lambda x = 0\}$ and $\mathrm{Im}\Lambda = \{y \in Y \mid y = \Lambda x \text{ for some } x \in X\}$. Also, $\Lambda^* : Y^* \to X^*$ denotes the adjoint of Λ, where X^* and Y^* denote the dual spaces of X and Y, respectively. By $\langle \lambda, y \rangle$ we denote the value of the functional $\lambda : Y^* \to X^*$ at $y \in Y$.

Let p be a natural number and let $B : X \times X \times \ldots \times X$ (with p copies of X) $\to$ Y be a continuous symmetric p-multilinear mapping. The p-form associated to B is the map $B[\cdot]^p : X \to Y$ defined by $B[x]^p = B(x, x, \ldots, x)$, for $x \in X$. If $F : X \to Y$ is a differentiable mapping, its Fréchet derivative at a point $x \in X$ will be denoted by $F'(x)$. If $F : X \to Y$ is of class C^p, we let $F^{(p)}(x)$ be the pth derivative of F at the point x (a symmetric multilinear map of p copies of X to Y) and the associated p-form, also called the pth–order mapping, is $F^{(p)}(x)[h]^p = F^{(p)}(x)(h, h, \ldots, h)$. Furthermore, we use the following key notation,

$$\mathrm{Ker}^p F^{(p)}(x) = \{h \in X \mid F^{(p)}(x)[h]^p = 0\}$$

is the p-kernel of F.

2 The p-Factor-Operator and p-Regular Mappings

In this section, we recall some definitions of the p–regularity theory [10,18].

Consider a sufficiently smooth mapping F from a Banach space X to a Banach space Y. The mapping F is called *regular* at some point $x^* \in X$ if $\mathrm{Im}\, F'(x^*) = Y$. We are interested in the case when the mapping F is *nonregular (irregular, degenerate)* at x^*, i.e., when $\mathrm{Im} F'(x^*) \neq Y$.

In this paper, we assume that $\operatorname{Im} F'(x^*)$ is closed. We recall a definition of the p-regular mapping and of the p-factor-operator. We construct a p-factor-operator under the assumption that the space Y is decomposed into a direct sum

$$Y = Y_1 \oplus \ldots \oplus Y_p, \tag{4}$$

where $Y_1 = \operatorname{Im} F'(x^*)$, $Y_i = \operatorname{cl\,span} (\operatorname{Im} P_{Z_i} F^{(i)}(x^*)[\cdot]^i)$, $i = 2, \ldots, p-1$, $Y_p = Z_p$, Z_i is a closed complementary subspace for $(Y_1 \oplus \ldots \oplus Y_{i-1})$ with respect to Y, $i = 2, \ldots, p$, and $P_{Z_i} : Y \to Z_i$ is the projection operator onto Z_i along $(Y_1 \oplus \ldots \oplus Y_{i-1})$ with respect to Y, $i = 2, \ldots, p$. Hence, $Z_i = Y_i \oplus \ldots \oplus Y_p$.

Define the mappings [10]

$$f_i(x) : X \to Y_i, \quad f_i(x) = P_{Y_i} F(x), \quad i = 1, \ldots, p, \tag{5}$$

where $P_{Y_i} : Y \to Y_i$ is the projection operator onto Y_i along $(Y_1 \oplus \ldots \oplus Y_{i-1} \oplus Y_{i+1} \oplus \ldots \oplus Y_p)$ with respect to Y, $i = 1, \ldots, p$.

The p-factor-operator plays the central role in the p-regularity theory. The number p is chosen as the minimum number for which (4) holds. We give the following definition of the p-factor-operator.

Definition 3. *The linear operator* $\bar{\Psi}_p(h) \in \mathcal{L}(X, Y_1 \oplus \ldots \oplus Y_p)$, *defined by*

$$\bar{\Psi}_p(h) = f_1'(x^*) + f_2''(x^*)[h] + \ldots + f_p^{(p)}(x^*)[h]^{p-1}, \quad h \in X,$$

is called a p-factor-operator of the mapping $F(x)$ at the point x^*.

We also introduce the corresponding inverse multivalued operator $\{\bar{\Psi}_p(h)\}^{-1}$ defined by

$$\{\bar{\Psi}_p(h)\}^{-1}(y) = \left\{ \xi \in X \mid \bar{\Psi}_p(h)[\xi] = f_1'(x^*)[\xi] + \ldots + f_p^{(p)}(x^*)[h^{p-1}, \xi] = y \right\}, \quad y \in Y.$$

In addition, $\|\{\bar{\Psi}_p(h)\}^{-1}\|$ is defined by

$$\|\{\bar{\Psi}_p(h)\}^{-1}\| = \sup_{\|y\|=1} \inf \{\|\xi\| \mid \bar{\Psi}_p(h)[\xi] = y\}.$$

Now we are ready to introduce another very important definition in the p-regularity theory.

Definition 4. *We say that the mapping F is p–regular at x^* along an element h if* $\operatorname{Im} \bar{\Psi}_p(h) = Y$.

A set $H_p(x^*)$ introduced below is the key one for many results obtained in the p-regularity theory.

$$H_p(x^*) = \bigcap_{i=1}^p \operatorname{Ker}^i f_i^{(i)}(x^*) = \left\{ h \in X \mid f_i^{(i)}(x^*)[h]^i = 0, \ i = 1, \ldots, p \right\}. \tag{6}$$

We are also using the set

$$H_p^\alpha(x^*) = \left\{ h \in X \mid \left\| f_i^{(i)}(x^*)[h]^i \right\|_{Y_i} \leq \alpha, \ \|h\|_X = 1, \ i = 1, \ldots, p \right\}, \tag{7}$$

where $\alpha > 0$ is some number.

Definition 5. *We say that the mapping F is strongly p–regular at x^* if there exists $\alpha > 0$ such that*

$$\sup_{h \in H_p^\alpha(x^*)} \|\{\bar{\Psi}_p(h)\}^{-1}\| < \infty.$$

We will need the following theorems to prove our results.

Theorem 2 ([1])**.** *Let X and Y be Banach spaces and U be a neighborhood of x^* in X. Assume that the mapping $F : U \to Y$ is Fréchet differentiable and is regular at x^*, i.e. $\operatorname{Im} F'(x^*) = Y$. Then there exists $U' \subset U$, $C > 0$ and the mapping $\xi \to x(\xi)$ such that $F(\xi + x(\xi)) = F(x^*)$ for any $\xi \in U'$ and*

$$\|x(\xi)\|_X \le C\|F(\xi) - F(x^*)\| \quad \forall \xi \in U'. \tag{8}$$

Theorem 3 ([10])**.** *Let X and Y be Banach spaces and U be a neighborhood of x^* in X. Assume that the mapping $F : U \to Y$ is p-times continuously differentiable at x^*, and F is strongly p-regular at x^*. Then there exists $U' \subset U$, $\delta > 0$ and the mapping $\xi \to x(\xi)$ such that $F(\xi + x(\xi)) = F(x^*)$ for any $\xi \in U'$ and*

$$\|x(\xi)\|_X \le \delta \sum_{i=1}^p \|f_i(\xi) - f_i(x^*)\|_{Y_i}^{1/i}, \quad \forall \xi \in U'. \tag{9}$$

3 Degenerate Optimization Problems with Equality Constraints

In this section, we analyze properties of the mapping $\mathcal{L}$ defined in (3), introduce a modified Lagrange system and derive generalized necessary conditions. Throughout this section we assume that $x \in X$ and $(\lambda_0, \lambda) \in (\mathbb{R} \times Y^*)\backslash\{0\}$ is a generalized Lagrange multiplier.

Let $Z = X \times \mathbb{R} \times Y^*$, $W = X^* \times Y \times \mathbb{R}$, and $z = (x, \lambda_0, \lambda)$, $z \in Z$. Then $\mathcal{L}$ defined by (3) is a mapping from space Z to space W. We assume that similarly to (4), the space W is decomposed into the direct sum:

$$W = W_1 \oplus \ldots \oplus W_p, \tag{10}$$

where $W_1 = \operatorname{Im} \mathcal{L}'(z^*)$, $W_i = \operatorname{cl\,span}(\operatorname{Im} P_{V_i} \mathcal{L}^{(i)}(z^*)[\cdot]^i)$, $i = 2, \ldots, p-1$, $W_p = V_p$, V_i is a closed complementary subspace for $(W_1 \oplus \ldots \oplus W_{i-1})$ with respect to W, $i = 2, \ldots, p$, and $P_{V_i} : W \to V_i$ is the projection operator onto V_i along $(W_1 \oplus \ldots \oplus W_{i-1})$ with respect to W, $i = 2, \ldots, p$. Hence, $V_i = W_i \oplus \ldots \oplus W_p$. The number p is chosen as the minimum number for which (10) holds.

Let $l_i(z)$ be the mappings associated with $\mathcal{L}(z)$ and defined similarly to (5) by

$$l_i(z) : Z \to W_i, \quad l_i(z) = P_i \mathcal{L}(z), \quad i = 1, \ldots, p, \tag{11}$$

where $P_i : W \to W_i$ is the projection operator onto W_i along $(W_1 \oplus \ldots \oplus W_{i-1} \oplus W_{i+1} \ldots \oplus W_p)$ with respect to W, $i = 1, \ldots, p$. Note that the definitions of P_i and $l_i(z)$ imply

$$l_i^{(r)}(z^*) = 0, \quad r = 1, \ldots, i-1, \quad i = 2, \ldots p. \tag{12}$$

We also define $\Psi_p(h) \in \mathcal{L}(Z, W_1 \oplus \ldots \oplus W_p)$ to be a p-factor-operator of the mapping $\mathcal{L}(z)$ at the point $z^* = (x^*, \lambda_0^*, \lambda^*)$,

$$\Psi_p(h) = l_1'(z^*) + l_2''(z^*)[h] + \ldots + l_p^{(p)}(z^*)[h]^{p-1}, \quad h \in Z. \tag{13}$$

Note that the kernel of the linear operator $\Psi_p(h)$ is defined as

$$\mathrm{Ker}\Psi_p(h) = \{\eta \in Z \mid l_i^{(i)}(z^*)[h^{i-1}, \eta] = 0, \quad i = 1, \ldots, p\}. \tag{14}$$

Also observe that by definition of l_i given in (11) and the definition of W_i, $i = 1, \ldots, p$, the equality

$$l_1'(z_k^*)[h] + \ldots + l_i^{(i)}(z_k^*)[h]^i + \ldots + l_p^{(p)}(z_k^*)[h]^p = 0 \tag{15}$$

holds for some $h \in X$ if and only if

$$l_i^{(i)}(z_k^*)[h]^i = 0, \quad i = 1, \ldots, p. \tag{16}$$

Introduce a *modified Lagrange system* as follows:

$$\bar{\mathcal{L}}(z) = l_1(z) + l_2'(z)[h] + \ldots + l_p^{(p-1)}(z)[h]^{p-1} = 0, \tag{17}$$

where mappings l_i are defined by (11), $i = 1, \ldots, p$. In the regular case, when the derivative of $\mathcal{L}(z^*)$, which is defined by (3), is onto, system (17) reduces to Lagrange system (3).

The construction of a p-factor-operator for mapping $\mathcal{L}(z)$ and introduction of the modified mapping $\bar{\mathcal{L}}(z)$ can be viewed as finding p–expansions of all equations in system (3). Since we treat $\mathcal{L}(z)$ as one mapping, all equations in (3) are expanded up to the p-th order. Note that even higher–order derivatives of the third equation in (3) are equal to zero, the derivatives up to the second order will contribute into the p-factor-operator.

Now we are ready to state and prove generalized necessary conditions for problem (1). Under an assumption that x^* is a solution of (1), the following theorem introduces conditions that guarantee that point $z^* = (x^*, \lambda_0^*, \lambda^*)$ is a regular solution of the modified Lagrange system $\bar{\mathcal{L}}(z) = 0$ defined in (17), and states additional assumptions under which the solution z^* is locally unique. A special version of the following theorem was given as a theorem in [20] without a proof. Note that Theorem 4 subsumes the case of $\lambda_0^* = 0$.

Theorem 4. *Let X and Y be Banach spaces, and let x^* be a solution of (1). Suppose that mapping $F : X \to Y$ and function $f : X \to \mathbb{R}$ are $(p + 1)$–times continuously Fréchet differentiable. Let (λ_0^*, λ^*) be a generalized Lagrange multiplier such that $z^* = (x^*, \lambda_0^*, \lambda^*)$ is a solution of (3). Then the following holds:*

1. *For any $h \in Z = X \times \mathbb{R} \times Y^*$, the point z^* is a solution of system (17).*
2. *If mapping $\mathcal{L} : Z \to W$ defined in (3) is p-regular at the point $z^* = (x^*, \lambda_0^*, \lambda^*)$ with respect to a vector h, then z^* is a regular solution of system (17).*

3. If the mapping $\mathcal{L}$ is p-regular at the point $z^* = (x^*, \lambda_0^*, \lambda^*)$ with respect to some vector h and the set $\mathrm{Ker}\Psi_p(h)$ defined by (14) consists only of the zero element, then z^* is a locally unique regular solution of system (17).

Proof. We give the proof in three parts corresponding to three statements of the theorem.

1. By an assumption of the theorem, $z^* = (x^*, \lambda_0^*, \lambda^*)$ is a solution of (3), that is $\mathcal{L}(z^*) = 0$, and, hence, $l_1(z^*) = 0$. Then by (12), $\bar{\mathcal{L}}(z^*) = 0$.
2. If mapping $\mathcal{L} : Z \to W$ is p-regular at the point $z^* = (x^*, \lambda_0^*, \lambda^*)$ with respect to some h, then $\mathrm{Im}\Psi_p(h) = W$, where $\Psi_p(h)$ is defined by (13). Then $\bar{\mathcal{L}}'(z^*) = \Psi_p(h)$ implies that $\bar{\mathcal{L}}$ is regular at z^*.
3. In this part, we assume that the set $\mathrm{Ker}\Psi_p(h)$ defined by (14) consists only of the zero element. By the second statement of the theorem, the point z^* is a regular solution of (17). Assume on the contrary that z^* is not locally unique. This implies existence of a sequence $\{z_k\} \to z^*$, $k \to \infty$, such that $\bar{\mathcal{L}}(z_k) = 0$. Let for $k \in \mathbb{N}$, $h_k = (z_k - z^*)/\|z_k - z^*\|$ and $t_k = \|z_k - z^*\|$. Then $z_k = z^* + t_k h_k$ and $\|h_k\| = 1$.
   By the Taylor expansion,

   $$\bar{\mathcal{L}}(z_k) = \bar{\mathcal{L}}(z^*) + \bar{\mathcal{L}}'(z^*)[t_k h_k] + \omega(t_k),$$

   where $\|\omega(t_k)\| = o(t_k)$. Since $\bar{\mathcal{L}}(z^*) = 0$ and $\bar{\mathcal{L}}(z_k) = 0$ then by (17),

   $$\begin{aligned} 0 &= \bar{\mathcal{L}}'(z^*)[t_k h_k] + \omega(t_k) \\ &= l_1'(z^*)[t_k h_k] + l_2''(z^*)[h, t_k h_k] + \ldots + l_p^{(p)}(z^*)[h^{p-1}, t_k h_k] + \omega(t_k). \end{aligned} \tag{18}$$

   By the definition of $l_1(z), \ldots, l_p(z)$, Eq. (18) implies that

   $$\| l_r^{(r)}(z^*)[h^{r-1}, t_k h_k] \| = o(t_k), \qquad r = 1, \ldots, p, \tag{19}$$

   for all $k \geq \bar{K}$, where $\bar{K} \in \mathbb{N}$ is some number.
   To finish the proof, we will get a contradiction to the assumption $\mathrm{Ker}\Psi_p(h) = \{0\}$. Namely, we will show that there exists $\hat{\eta} \neq 0$ such that $\hat{\eta} \in \mathrm{Ker}\Psi_p(h)$. Consider the mapping

   $$\mathcal{L}_1(\eta) = l_1'(z^*)[\eta] + \ldots + l_i^{(i)}(z^*)[h^{i-1}, \eta] + \ldots + l_p^{(p)}(z^*)[h^{p-1}, \eta]. \tag{20}$$

   By comparing (14) and (20), we observe that $\mathrm{Ker}\Psi_p(h) = \{\eta \in Z \mid \mathcal{L}_1(\eta) = 0\}$. Hence, to get a contradiction to the assumption $\mathrm{Ker}\Psi_p(h) = \{0\}$, it is enough to construct a vector $\bar{\eta}$ such that $\mathcal{L}_1(\bar{\eta}) = 0$. We will verify that all conditions of Theorem 2 hold with $F = \mathcal{L}_1$ and $\bar{\eta} = 0$. By an assumption of the theorem, the mapping $\mathcal{L}(z)$, defined in (3), is p-regular at the point $z^* = (x^*, \lambda_0^*, \lambda^*)$ with respect to the vector $h \neq 0$, that is $\mathrm{Im}\Psi_p(h) = W$. Note that $\mathcal{L}_1'(0) = \Psi_p(h)$ and, hence, $\mathcal{L}_1$ is regular at $\bar{\eta} = 0$. By Theorem 2, there exists a neighborhood U' of 0 and a mapping $\xi \to x(\xi)$ such that $\mathcal{L}_1(\xi + x(\xi)) = \mathcal{L}_1(0) = 0$. Then for a sufficiently big k, we get that $\xi = t_k h_k \in U'$ and $\mathcal{L}_1(t_k h_k + x(t_k h_k)) = 0$. Note that $\|t_k h_k + x(t_k h_k)\| \neq 0$ because of (8), (19), and (20). Let $\hat{\eta} = t_k h_k + x(t_k h_k)$, then $\hat{\eta} \in \mathrm{Ker}\Psi_p(h)$, which is a contradiction. Thus, system (17) has a locally unique regular solution z^*.

The following theorem follows directly from Theorem 4 and can be considered as generalized necessary optimality conditions for problem (1).

Theorem 5. *Let X and Y be Banach spaces, and let x^* be a solution of (1). Suppose that the mapping $F : X \to Y$ and the function $f : X \to \mathbb{R}$ are $(p+1)$–times continuously Fréchet differentiable. Then there exists a generalized Lagrange multiplier (λ_0^*, λ^*) such that:*

$$\bar{\mathcal{L}}(x^*, \lambda_0^*, \lambda^*) = 0, \tag{21}$$

where not all of λ_0^*, λ^* are zero and $\bar{\mathcal{L}}$ is defined in (17) with some $h \in Z$. Moreover, if the mapping $\mathcal{L}$ is p-regular at the point $z^* = (x^*, \lambda_0^*, \lambda^*)$ with respect to the vector $h \in Z$, then z^* is a regular solution of (21). If, in addition, $\mathrm{Ker}\Psi_p(h) = \{0\}$, where $\mathrm{Ker}\Psi_p(h)$ is defined in (14), then z^* is a locally unique regular solution of system (21).

Proof. The classical first-order Euler-Lagrange necessary optimality conditions [1] imply existence of a generalized Lagrange multiplier (λ_0^*, λ^*) such that $z^* = (x^*, \lambda_0^*, \lambda^*)$ is a solution of (3). Then the statement of the theorem follows from Theorem 4.

Remark 1. If $(\lambda_0^*, \lambda^*) = (0,0)$, then (2) is satisfied trivially. Note that both the classical necessary conditions [1] as well as the third equation of system (3) require $(\lambda_0^*, \lambda^*) \neq (0,0)$.

The following theorem gives a simplified form of the modified Lagrange system in the specific case of $p = 2$ under an additional assumption.

Theorem 6. *Let X and Y be Banach spaces, and let x^* be a solution of (1). Suppose that mapping $F : X \to Y$ and function $f : X \to \mathbb{R}$ are twice continuously Fréchet differentiable. Let (λ_0^*, λ^*) be a generalized Lagrange multiplier such that $z^* = (x^*, \lambda_0^*, \lambda^*)$ is a solution of (3). Assume that $\mathcal{L}'(z^*)h = 0$ for some $h \neq 0$. Then the point z^* is a solution of the following system:*

$$\tilde{\mathcal{L}}(z) = \mathcal{L}(z) + \mathcal{L}'(z)h = 0. \tag{22}$$

If, in addition, $\mathrm{Im}\tilde{\mathcal{L}}'(z^*)$ is onto and $\mathrm{Ker}\tilde{\mathcal{L}}'(z^*) = \{0\}$, then z^* is a locally unique regular solution of system (22).

Proof. Proof is similar to the proof of Theorem 4.

Remark 2. Theorem 5 can also be stated and proved in the case of $p = 2$ for the modified system (22) under an additional assumption that $\mathcal{L}'(z^*)h = 0$ for some $h \neq 0$.

Now, we turn our attention to the classical Lagrange system (3). The following theorem gives new conditions under which Lagrange system (3) has a locally unique solution.

Similarly to (6) and (7), we define the following sets associated with the mappings l_i:

$$H_p(z^*) = \{h \in Z \mid l_i^{(i)}(z^*)[h]^i = 0, \quad i = 1, \ldots, p\}, \tag{23}$$

and

$$H_p^\alpha(z^*) = \{h \in Z \mid \left\| l_i^{(i)}(z^*)[h]^i \right\|_{W_i} \leq \alpha, \|h\|_Z = 1, \ i = 1, \ldots, p\}, \tag{24}$$

where $\alpha > 0$.

Theorem 7. *Let X and Y be Banach spaces, and let x^* be a solution of (1). Suppose that the mapping $F : X \to Y$ and the function $f : X \to \mathbb{R}$ are $(p+1)$-times continuously Fréchet differentiable. Let (λ_0^*, λ^*) be a generalized Lagrange multiplier such that $(x^*, \lambda_0^*, \lambda^*)$ is a solution of (3). Assume that the mapping $\mathcal{L}(z)$, defined in (3), is strongly p-regular at the point $z^* = (x^*, \lambda_0^*, \lambda^*)$. Assume also that $H_p(z^*) = \{0\}$, where $H_p(z^*)$ is defined in (23). Then system (3) has a locally unique solution z^*.*

Proof. By an assumption of the theorem, $z^* = (x^*, \lambda_0^*, \lambda^*)$ is a solution of (3) with some generalized Lagrange multiplier (λ_0^*, λ^*). Assume on the contrary that z^* is not a locally unique solution of (3). This implies existence of a sequence $\{z_k\} \to z^*$, $k \to \infty$, such that $\mathcal{L}(z_k) = 0$. Let $h_k = (z_k - z^*)/\|z_k - z^*\|$ for all $k \in \mathbb{N}$. By the Taylor expansion,

$$\mathcal{L}(z_k) = \mathcal{L}(z^*) + \mathcal{L}'(z^*)(z_k - z^*) + \ldots + \frac{1}{p!}\mathcal{L}^{(p)}(z^*)(z_k - z^*)^p + \omega(z_k),$$

where $\omega(z_k) = o(\|z_k - z^*\|^p)$, or

$$0 = \mathcal{L}'(z^*)(z_k - z^*) + \ldots + \frac{1}{p!}\mathcal{L}^{(p)}(z^*)(z_k - z^*)^p + \omega(z_k). \tag{25}$$

Recall that $P_r : W \to W_r$ is the projection operator introduced in (11). Then, for any $r = 1, \ldots, p$, dividing the equality (25) by $\|z_k - z^*\|^r$, multiplying by P_r and taking limits as $k \to \infty$, we get that there exists $K \in \mathbb{N}$ such that for any $k \geq K$, $\|P_r \mathcal{L}^{(r)}(z^*)[h_k]^r\| \leq \alpha$, $r = 1, \ldots, p$, where α is a sufficiently small number. By the definition of $l_r(z)$, the last inequality implies

$$\|l_r^{(r)}(z^*)[h_k]^r \| \leq \alpha, \qquad r = 1, \ldots, p, \tag{26}$$

for a sufficiently small α and for all $k \geq K$. Hence,

$$h_k \in H_p^\alpha(z^*), \quad k \geq K,$$

where set $H_p^\alpha(z^*)$ is defined by (24).

To finish the proof, we will get a contradiction to the assumption $H_p(z^*) = \{0\}$. Namely, we will show that there exists $\hat{h} \neq 0$ such that $\hat{h} \in H_p(z^*)$.

Consider the mapping

$$\tilde{L}(h) = l_1'(z^*)[h] + \ldots + l_i^{(i)}(z^*)[h]^i + \ldots + l_p^{(p)}(z^*)[h]^p, \quad h \in Z. \qquad (27)$$

By comparing (23) with (27) and using (15)–(16), we observe that $H_p(z^*) = \{h \in Z \mid \tilde{L}(h) = 0\}$ and $\tilde{L}(0) = 0$. Hence, to get a contradiction to the assumption $H_p(z^*) = \{0\}$, it is enough to construct a vector $\hat{h}$ such that $\tilde{L}(\hat{h}) = 0$.

We will verify that all conditions of Theorem 3 hold with $F = \tilde{L}$ and $z^* = 0$. Note that mappings $\tilde{L}(z)$ and $\mathcal{L}(z)$ have the same p-factor-operator at the point z^* and the same corresponding set $H_p^\alpha(z^*)$. This implies that $\tilde{L}(z)$ is strongly p-regular at the point z^*. By Theorem 3, there exists a neighborhood U' of 0 and a mapping $\xi \to x(\xi)$ such that $\tilde{L}(\xi + x(\xi)) = \tilde{L}(0) = 0$. Then for a sufficiently big k and $t = \|z_k - z^*\|$, we get $\xi = th_k \in U'$ and $\tilde{L}(th_k + x(th_k)) = 0$. Note that $\|th_k + x(th_k)\| \neq 0$ because of (9), (26), and (27). Let $\hat{h} = th_k + x(th_k)$ then $\hat{h} \in H_p(z^*)$, which is a contradiction. Thus, system (3) has a locally unique solution z^*.

4 Examples

Example 1. This example was mentioned in Sect. 1. It illustrates the situation when the problem does not satisfy classical necessary optimality conditions (2) with $\lambda_0^* \neq 0$. However, Theorem 4 derived in the paper can be applied to the example.

Consider the problem

$$\underset{(x_1,x_2)}{\text{minimize}} \ x_2 \quad \text{subject to} \quad F(x_1,x_2) = x_1^2 - x_2^3 = 0, \qquad (28)$$

which has a solution $(x_1^*, x_2^*)^T = (0,0)^T$. Note that $F'(0,0) = (0,0)$. The Lagrangian for (28) is $L(x_1, x_2, \lambda_0, \lambda) = \lambda_0 x_2 + (x_1^2 - x_2^3)\lambda$, and the Lagrange system (3) has the form

$$\mathcal{L}(x_1, x_2, \lambda_0, \lambda) = \begin{pmatrix} 2x_1\lambda \\ \lambda_0 - 3x_2^2\lambda \\ x_1^2 - x_2^3 \\ \lambda_0^2 + \lambda^2 - 1 \end{pmatrix} = 0. \qquad (29)$$

System (29) has a unique degenerate solution $(x_1^*, x_2^*, \lambda_0, \lambda^*) = (0,0,0,1)$. Note that system (29) does not have a solution with a regular Lagrange multiplier when $\lambda_0 \neq 0$.

We will construct the modified Lagrange system (17) for problem (28).

In this example, $z^* = (0,0,0,1)$, $Y_1 = \text{Im}\,\mathcal{L}'(z^*) = \text{span}\,((1,0,0,0)^T, (0,1,0,0)^T, (0,0,0,1)^T)$, $Z_2 = Z_3 = \text{span}\,((0,0,1,0)^T)$ and

$$P = P_{Z_i} = \begin{pmatrix} 0 & 0 & 0 & 0 \\ 0 & 0 & 0 & 0 \\ 0 & 0 & 1 & 0 \\ 0 & 0 & 0 & 0 \end{pmatrix}, \quad i = 2,3.$$

In addition, $Y_2 = \mathrm{Im} P\mathcal{L}''(z^*)[\cdot]^2 = 0$ and $Y_3 = Z_3$. One can verify that $\mathcal{L}$ is 3-regular at z^* with respect to the vector $h = (0,1,0,0)^T$. Hence, with $z = (x_1, x_2, \lambda_0, \lambda)$,

$$l_1(z) = P_{Y_1}\mathcal{L}(z) = \begin{pmatrix} 2x_1\lambda \\ \lambda_0 - 3x_2^2\lambda \\ 0 \\ \lambda_0^2 + \lambda^2 - 1 \end{pmatrix}, \quad l_2(z) = 0, \quad l_3(z) = P_{Y_3}\mathcal{L}(z) = \begin{pmatrix} 0 \\ 0 \\ x_1^2 - x_2^3 \\ 0 \end{pmatrix}.$$

Then, by (17) with $h = (0,1,0,0)^T$, the modified Lagrange system is

$$\bar{\mathcal{L}}(z) = l_1(z) + l_3''(z)[h]^2 = \begin{pmatrix} 2x_1\lambda \\ \lambda_0 - 3x_2^2\lambda \\ -6x_2 \\ \lambda_0^2 + \lambda^2 - 1 \end{pmatrix} = 0. \tag{30}$$

Note that neither 2−regularity assumption nor strong 3−regularity assumption holds and hence, the assumptions of Theorem 7 are not satisfied. However, the assumptions of Theorem 4 and Theorem 5 hold. By the third statement in Theorem 4, we get that system (30) has a locally unique regular solution $(0,0,0,1)^T$. □

Example 2. This example illustrates application of Theorem 5 to a calculus of variations problem. Consider the problem of minimizing the functional

$$J_0[x] = \int_0^\pi x^2(\tau)d\tau \tag{31}$$

over $x \in C^2([0,\pi])$ subject to the constraints

$$x(0) = x(\pi) = 0, \quad J_1[x] = \int_0^\pi \left(x^2(\tau) - (x'(\tau))^2\right)d\tau = 0. \tag{32}$$

The solution in this example is $x^* \equiv 0$.

In more general setting, problem (31)–(32) is a special case of the problem of minimizing the functional $J_0 = \int_{t_1}^{t_2} F(t,x,x')dt$ subject to the constraint

$$J_1[x] = \int_{t_1}^{t_2} G(t,x,x')dt = l$$

and boundary conditions

$$x(t_1) = x_1, \quad x(t_2) = x_2,$$

where $x \in C^2([t_1,t_2], \mathbb{R})$ and $F(t,x,x')$ is a function with continuous first and second derivatives with respect to its arguments.

According to [22], if a function $x(\cdot)$ yields a local extremum (relative to the space $C^2([t_1, t_2])$), then there are numbers, λ_0 and λ_1, not vanishing simultaneously, such that the Euler equation holds:

$$\frac{d}{dt}(\lambda_0 F_{x'} + \lambda G_{x'}) = \lambda_0 F_x + \lambda G_x. \tag{33}$$

Notice that Eq. (33) is just the Euler-Lagrange equation for the functional $\lambda_0 J_0 + \lambda J_1$.

For problem (31)–(32) in this example, (33) implies that if x^* is an extremum, then

$$\lambda_0 x^* + \lambda x^* + \lambda x^{*\prime\prime} = 0.$$

For the solution $x^* \equiv 0$, the last equality is satisfied with any (λ_0^*, λ^*) including $(\lambda_0^*, \lambda^*) = (0, 1)$. For problem (31)–(32), the system of optimality conditions can be written as

$$\mathcal{L}(x, \lambda_0, \lambda) = \begin{pmatrix} \lambda_0 x + \lambda x + \lambda x'' \\ \int_0^\pi \left(x^2(\tau) - (x'(\tau))^2 \right) d\tau \\ \lambda_0^2 + \lambda^2 - 1 \end{pmatrix} = 0. \tag{34}$$

System (34) has a family of solutions $x = a \sin t$ with $\lambda = (\lambda_0^*, \lambda^*) = (0, 1)$ in a neighborhood of $x = 0$ including $x = 0$ with $a = 0$. Note that differentiability properties of $\int_0^\pi \left(x^2(\tau) - (x'(\tau))^2 \right) d\tau$ are described, for example, in [1], [10], and [22].

The derivative of $\mathcal{L}$ defined in (34) is given by

$$\mathcal{L}'(x, \lambda_0, \lambda) \begin{pmatrix} x^* \\ \lambda_0^* \\ \lambda^* \end{pmatrix} = \begin{pmatrix} (\lambda_0 + \lambda)x^* + \lambda x^{*\prime\prime} + \lambda_0^* x + \lambda^*(x + x'') \\ \int_0^\pi (2xx^* - 2x'x^{*\prime}) d\tau \\ 2\lambda_0 \lambda_0^* + 2\lambda \lambda^* \end{pmatrix}.$$

As is evident, $\mathcal{L}(x, \lambda_0, \lambda)$ is degenerate at $(0, 0, 1)$; indeed,

$$\mathcal{L}'(0, 0, 1) = \begin{pmatrix} (\cdot) + (\cdot)'' & 0 & 0 \\ 0 & 0 & 0 \\ 0 & 0 & 2 \end{pmatrix}.$$

The modified system of optimality conditions (22) for problem (31)–(32) has the form

$$\bar{\mathcal{L}}(x, \lambda_0, \lambda) = \mathcal{L}(x, \lambda_0, \lambda) + \mathcal{L}'(x, \lambda_0, \lambda)h(t)$$

$$= \begin{pmatrix} \lambda_0 x + \lambda x + \lambda x'' + (\lambda_0 + \lambda)(\sin 2t) - 4\lambda(\sin 2t) + x \\ \int_0^\pi \left(x^2(\tau) - (x'(\tau))^2 + 2x(\tau)(\sin 2\tau) - 2x'(2\cos 2\tau) \right) d\tau \\ \lambda_0^2 + \lambda^2 - 1 + 2\lambda_0 \end{pmatrix} = 0, \tag{35}$$

where $h(t) = (\sin 2t, 1, 0)$. Then, for any $(x^*, \lambda_0^*, \lambda^*)^T$,

$$\bar{\mathcal{L}}'(0,0,1) \begin{pmatrix} x^* \\ \lambda_0^* \\ \lambda^* \end{pmatrix} = \begin{pmatrix} 2(x^*) + (x^*)'' + (\lambda_0^* - 3\lambda^*) \sin 2t \\ \int\limits_0^\pi (2(x^*)(\sin 2\tau) - 4(x^*)'(\cos 2\tau)) \, d\tau \\ 2(\lambda_0^* + \lambda^*) \end{pmatrix}.$$

Hence, $\bar{\mathcal{L}}'(0,0,1)$ is surjective and the mapping $\bar{\mathcal{L}}(x, \lambda_0, \lambda)$ is nonsingular at $(0,0,1)$. System (35) has the point $(x^*, \lambda_0^*, \lambda^*) = (0,0,1)$ as its regular solution. This solution is locally unique by Theorem 5 since $\mathrm{Ker}\bar{\mathcal{L}}'(0,0,1) = \{0\}$.   □

Example 3. This example illustrates application of Theorem 5 and Theorem 7. Consider the problem

$$\underset{x}{\text{minimize}} \; x \quad \text{subject to} \quad F(x) = x^p = 0, \tag{36}$$

where $x \in \mathbb{R}$. This problem has a solution $x^* = 0$ and $F'(0) = 0$. The Lagrangian for (36) is $L(x, \lambda_0, \lambda) = \lambda_0 x + \lambda x^p$. Then Lagrange system (3) for problem (36) has the form

$$\mathcal{L}(x, \lambda_0, \lambda) = \begin{pmatrix} \lambda_0 + p\lambda x^{p-1} \\ x^p \\ \lambda_0^2 + \lambda^2 - 1 \end{pmatrix} = 0. \tag{37}$$

First, note that system (37) does not have a solution with $\lambda_0 \neq 0$. Note also that the Jacobian of (37) is given by

$$J(x, \lambda_0, \lambda) = \mathcal{L}'(x, \lambda_0, \lambda) = \begin{pmatrix} p(p-1)\lambda x^{p-2} & 1 & px^{p-1} \\ px^{p-1} & 0 & 0 \\ 0 & 2\lambda_0 & 2\lambda \end{pmatrix}$$

and

$$J(0,0,1) = \begin{pmatrix} 0 & 1 & 0 \\ 0 & 0 & 0 \\ 0 & 0 & 2 \end{pmatrix}.$$

Hence, the Jacobian $J(0,0,1)$ is singular and system (37) has a degenerate solution $z^* = (0,0,1)$.

The corresponding to $\mathcal{L}$ p-factor-operator $\Psi(h)$ with $h = (1,0,0)$ and $z^* = (0,0,1)$ is defined by

$$\Psi(h) = l_1'(z^*) + \ldots + l_p^{(p)}(z^*)[h]^{p-1} = \begin{pmatrix} 0 & 1 & 0 \\ p! & 0 & 0 \\ 0 & 0 & 2 \end{pmatrix}$$

and satisfies $\mathrm{Im}\Psi_p(h) = \mathbb{R}^3$. Hence, $\mathcal{L}$ defined by (37) is p-regular at z^* with respect to the vector h. In addition, $\mathcal{L}(x, \lambda_0, \lambda)$ is uniformly strongly p-regular at $(0,0,1)^T$ and the corresponding to $\mathcal{L}$ set is $H_p(0,0,1) = \{0\}$. Then Theorem 7 guarantees that system (37) has a locally unique solution.

The modified Lagrange system in this example is

$$\bar{\mathcal{L}}(x, \lambda_0, \lambda) = \begin{pmatrix} \lambda_0 + p\lambda x^{p-1} \\ p!x \\ \lambda_0^2 + \lambda^2 - 1 \end{pmatrix} = 0.$$

By Theorem 5, this system has a locally unique regular solution $(0, 0, 1)^T$.   □

Example 4. This example illustrates the case when an objective function is present explicitly with a nonzero multiplier in a modified Lagrange system. Consider the problem

$$\begin{array}{c} \text{minimize } x_1^2 \\ {\scriptstyle (x_1, x_2)} \\ \text{subject to } x_1 x_2 = 0, \\ x_2^2 = 0, \end{array} \tag{38}$$

which has a solution $(x_1^*, x_2^*)^T = (0, 0)^T$. Note that $F'(0, 0) = 0$. Lagrange system (3) for problem (38) is

$$\mathcal{L}(x_1, x_2, \lambda_0, \lambda_1, \lambda_2) = \begin{pmatrix} 2\lambda_0 x_1 + \lambda_1 x_2 \\ \lambda_1 x_1 + 2\lambda_2 x_2 \\ x_1 x_2 \\ x_2^2 \\ \lambda_0^2 + \lambda_1^2 + \lambda_2^2 - 1 \end{pmatrix} = 0. \tag{39}$$

System (39) has a degenerate solution $(x_1^*, x_2^*, \lambda_0^*, \lambda_1^*, \lambda_2^*) = (0, 0, 0, 0, 1)$, which is not locally unique.

The modified Lagrange system with $h = (1, 1, 1, 0, 0)^T$ has the form

$$\bar{\mathcal{L}}(x_1, x_2, \lambda_0, \lambda_1, \lambda_2) = \begin{pmatrix} 2\lambda_0 + \lambda_1 + 2x_1 \\ \lambda_1 x_1 + 2\lambda_2 x_2 \\ x_1 + x_2 \\ 2x_2 \\ \lambda_0^2 + \lambda_1^2 + \lambda_2^2 - 1 \end{pmatrix} = 0. \tag{40}$$

Note that the term $2x_1$ is the derivative of the objective function, which is present explicitly in the first equation with a nonzero coefficient. Moreover, the point $(x_1^*, 0, 0, 0, 1)$ is a solution of the optimality system (39) with any x_1^* while optimization problem (38) has a unique solution $(x_1^*, x_2^*)^T = (0, 0)^T$. In contrast to the classical Lagrange system (39), the modified system (40) yields $(x_1^*, x_2^*)^T = (0, 0)^T$.

If the vector h is chosen in such a way that $\bar{\mathcal{L}}'(0, 0, 0, 0, 1)$ is regular, then by Theorem 5 system (40) has the locally unique solution.   □

References

1. Ioffe, A.D., Tihomirov, V.M.: Theory of Extremal Problems. North-Holland, Amsterdam (1979)

2. Brezhneva, O.A., Tret'yakov, A.A.: Optimality conditions for degenerate extremum problems with equality constraints. SIAM J. Control. Optim. **42**, 729–745 (2003)
3. Brezhneva, O., Tret'yakov, A.: The p-th order necessary optimality conditions for inequality—constrained optimization problems. In: Kumar, V., Gavrilova, M.L., Tan, C.J.K., L'Ecuyer, P. (eds.) ICCSA 2003. LNCS, vol. 2667, pp. 903–911. Springer, Heidelberg (2003). https://doi.org/10.1007/3-540-44839-X_95
4. Brezhneva, O.A., Tret'yakov, A.A.: The pth order optimality conditions for inequality constrained optimization problems. Nonlinear Anal. **63**, e1357–e1366 (2005)
5. Brezhneva, O.A., Tret'yakov, A.A.: The pth order optimality conditions for non-regular optimization problems. Dokl. Math. **77**, 163–165 (2008)
6. Dmitruk, A.V.: Quadratic conditions for a Pontryagin minimum in an optimal control problem linear in the control. Theorems Weak. Equal. Constr. Math. USSR Izv. **31**, 121–141 (1986)
7. Izmailov, A.F.: Optimality conditions for degenerate extremum problems with inequality-type constraints. Comput. Math. Math Phys. **34**, 723–736 (1994)
8. Izmailov, A.F., Solodov, M.V.: Optimality conditions for irregular inequality-constrained problems. SIAM J. Control. Optim. **40**, 1280–1295 (2001)
9. Izmailov, A.F., Solodov, M.V.: The theory of 2-regularity for mappings with Lipschitzian derivatives and its applications to optimality conditions. Math. Oper. Res. **27**, 614–635 (2002)
10. Izmailov, A.F., Tret'yakov, A.A.: Factor-Analysis of Nonlinear Mappings. Nauka, Moscow (in Russian) (1994)
11. Ledzewicz, U., Schättler, H.: Second-order conditions for extremum problems with nonregular equality constraints. J. Optim. Theory Appl. **86**, 113–144 (1995)
12. Ledzewicz, U., Schättler, H.: A high-order generalization of the Lyusternik theorem. Nonlinear Anal. **34**, 793–815 (1998)
13. Ledzewicz, U., Schättler, H.: High-order approximations and generalized necessary conditions for optimality. SIAM J. Control. Optim. **37**, 33–53 (1999)
14. Milyutin, A.A.: On quadratic conditions for an extremum in smooth problems with a finite-dimensional image. In: Lenin, V.L. (ed.) Methods of the Theory of Extremal Problems in Economics, Nauka, Moscow, pp. 138–177 (in Russian) (1981)
15. Tret'yakov, A.A.: Necessary conditions for optimality of pth order. In: Control and Optimization, MSU, Moscow, pp. 28–35 (in Russian) (1983)
16. Tret'yakov, A.A.: Necessary and sufficient conditions for optimality of pth order. USSR Comput. Math. Math. Phys. **24**, 123–127 (1984)
17. Tret'yakov, A.A.: The implicit function theorem in degenerate problems. Russ. Math. Surv. **42**, 179–180 (1987)
18. Tret'yakov, A.A., Marsden, J.E.: Factor-analysis of nonlinear mappings: p-regularity theory. Commun. Pure Appl. Anal. **2**, 425–445 (2003)
19. Brezhneva, O.A., Tret'yakov, A.A.: The p-factor-Lagrange methods for degenerate nonlinear programming. Numer. Funct. Anal. Optim. **28**, 1051–1086 (2007)
20. Brezhneva, O.A., Tret'yakov, A.A.: P-factor-approach to degenerate optimization problems. In: Ceragioli, F., Dontchev, A., Futura, H., Marti, K., Pandolfi, L. (eds.) CSMO 2005. IFIP, vol. 199, pp. 83–90. Springer, Boston, MA (2006). https://doi.org/10.1007/0-387-33006-2_8
21. Rudin, W.: Functional Analysis. McGraw-Hill, Boston (1991)
22. Alekseev, V.M., Tikhomirov, V.M., Fomin, S.V.: Optimal Control. Consultants Bureau, New York and London (1987)

The Relative Formulation of the Quadratic Programming Problem in the Aircraft Assembly Modeling

Maria Stefanova[✉][iD] and Stanislav Baklanov[iD]

Peter the Great St.Petersburg Polytechnic University, St.Petersburg 195251, Russia
stefanova.m@list.ru

Abstract. Aircraft assembly modeling requires mass solving of quadratic programming (QP) problems. Thus, reducing the computation time becomes a very urgent task. Earlier [12], it was shown that a significant reduction in time can be achieved by reformulating the original QP problem in the so-called form of the relative QP problem. The current paper discusses the aspects of using relative QP problem related to the accuracy of the computations. Data preparation for the relative QP problem involves the matrix inversions. Since the Hessian in the considered QP problems is ill-conditioned, this operation leads to a rapid accumulation of round-off errors. The paper proposes to use the error-free operations of addition and multiplication to increase the accuracy of the QP problem solution and to maintain high computation speed for QP problem solving. Finally, a comparison of the primal, dual, and relative QP problem formulations in terms of computation time and result accuracy for assembly problems is presented.

Keywords: Quadratic programming · Relative problem formulation · Assembly problems · Computational accuracy

1 Introduction

Variation simulation analysis allows the estimation of the assembly quality, taking into account random deviations that occur during the mass production of parts. This field is most important for industries such as the automotive and aircraft manufacturing [4,15], which require significant costs, production time and high quality of the final product. Variation simulation analysis uses the Monte Carlo method to model the effects of each individual tolerance distribution and the complex relationships between these tolerances for the final assembly. With regard to the aviation industry, the safety and durability of products are important and determine high quality requirements. Thus, such an analysis should

The research was supported by Russian Science Foundation (project No. 22-19-00062, https://rscf.ru/en/project/22-19-00062/).

take into account the influence of deformations and contact interaction, the possibility of accounting the friction, dynamic load, the impact of sealant on the final assembly etc. In other words, one Monte Carlo simulation should include the solution of a nonlinear contact problem, which is the most time-consuming operation in the modeling of the assembly process. Time intensive computations, coupled with a large number of required simulations, obviously give rise to the problem of reducing the time of variation simulation analysis in general, and of one contact problem solving in particular. The authors of [5] propose to reduce the solution of contact problems to the solution of quadratic programming (QP) problems using optimization methods adapted to the specifics of assembly problems. This technique helps to significantly reduce the time of computations and even to carry out variation simulations on a personal computer in a reasonable time.

Some important aircraft joints, such as the wing to the fuselage, have a possible contact area between constituent parts much smaller than the size of the these parts. The reduction of the contact problem from the full-size finite element model to the area of possible contact only [9,13] leads to a significant reduction in dimension, as well as to a decrease in the computational time for solving the contact problem. The considered approach determines the structure of the Hessian matrix for the quadratic programming problem, that is block diagonal with dense blocks, symmetric and ill-conditioned. The Hessian matrix as well as the constraint matrix remain unchanged for all the simulations of the Monte Carlo method. This feature makes it possible to use optimization methods more efficiently by reformulating the original QP problem in a form that is more suitable in terms of computation time. Together with the primal and dual QP formulations, the QP problem may be formulated as a relative QP problem [12]. The relative QP problem has a simpler form of the constraint matrix and has a smaller dimension compared to the original QP problem. At the same time, all these problems are equivalent, in the sense that each solution can be easily obtained from the other.

However, the use of an approach based on relative QP formulation presented in [12], creates additional technical difficulties. The Hessian of the relative problem is obtained by the multiplication and inversion operations on the Hessian and constraint matrix of the original QP problem. The most troublesome is matrix inversion, that is based on the Cholesky decomposition. Due to the ill-conditioned Hessian of original QP problem, the multiple operations with matrix elements introduce the accumulation of rounding errors into the Hessian of the relative problem. Ultimately, the time of computations is decreased, while the accuracy of the contact problem solution obtained using the relative formulation is significantly reduced.

The current paper is devoted to the description of the modification of the relative QP problem approach for solving assembly problems in order to increase its accuracy while saving the computation time. Although this modification was used to obtain the results of computations in earlier works [1,12], it was not described. The modification is based on methods that improve the accuracy of operations for ill-conditioned matrices. Ogita [7,8,14] suggests using error-free arithmetic operations for summation and multiplication with floating point numbers while inverting matrix. The work [8] proposes iterative algorithm for accurate inverse Cholesky decomposition, which is further used to calculate the matrix inversions. A similar idea, based on the iterative repetition of matrix inversion with increasing accuracy of the result is used by Rump [11].

The paper is organized as follows. The Sect. 2 discusses the formulation of assembly problems in the form of different formulations of QP problems, the Sect. 3 describes error-free matrix operations and their implementation in assembly problems. In the Sect. 4, a modification of the relative QP problem approach, which preserves the high computation time and the required accuracy of the solution, is presented. A comparison of different formulations of the QP problem used to solve the assembly problems is presented in the Sect. 5 for a number of aircraft assembly models.

2   Assembly Problem Statement

Variation simulation analysis is used in assembly problems to evaluate the deviation of the final assembly from its nominal state based on the deviations of the parts of the joint. Since the deviations are random in nature, the variation simulation analysis requires a large number of simulations to obtain generalized characteristics of the final assembly, which are subsequently used for development the new or for modification the existing assembly technology. Each simulation includes the solution of a contact problem, which is formulated as a quadratic programming problem [5]:

$$\min_x P(x) = \min_x \tfrac{1}{2}x^T K x - f^T x, \\ \text{s.t. } A^T x - g \le 0, \tag{1}$$

where $x \in \mathbb{R}^n$ is the vector of geometrically restricted displacements in the area of possible contact $(n \ge m)$, $K \in \mathbb{R}^{n \times n}$ is the reduced stiffness matrix (positive definite block diagonal matrix with dense blocks), $A \in \mathbb{R}^{n \times m}$ is a linear operator that defines non penetration condition (full rank sparse matrix), $f \in \mathbb{R}^n$ is the vector of loads from fastening elements, $g \in \mathbb{R}^m$ is the vector of initial gap between parts.

The matrix A describes the topology of the joint, and the matrix K reflects the mechanical properties of the parts and the influence of the assembly jig. All these characteristics do not change during the assembly, so the matrices K and A do not change either. Deviation of parts are caused by inaccuracies in manufacturing and in positioning in the assembly jig, and are described by the initial gap vector g. The vector g varies from one joint instance to another. Arrangement of fasteners is included in the simulation through the vector of external loads f and can also vary, depending on the assembly technology. Thus, in general, variation simulation analysis requires to solve a set of p QP problems (1) (Fig. 1):

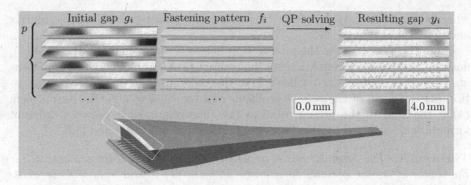

Fig. 1. The set of QP problems for variation simulation analysis.

$$\min_{x_i} P_i(x_i) = \min_{x_i} \tfrac{1}{2} x_i^T K x_i - f_i^T x_i, \quad (i = 1..p)$$
$$\text{s.t. } A^T x_i - g_i \leq 0, \tag{2}$$

where each problem is formulated for the one i-th of many possible assembly states, which depends on the initial gap between the parts $g_i \in \mathbb{R}^m$, the used fastening pattern $f_i \in \mathbb{R}^n$, and gives the displacement of parts after the assembly $x_i \in \mathbb{R}^n$ and resulting gap $y_i = g_i - A^T x_i$. Since the number of problems p is large ($10^3 - 10^6$), the main goal is to develop a fast tool for solving the set of problems (2). This is achieved using specially adapted optimization methods and reformulation of the original problem (1) [12].

The fact that the matrices K and A are usually the same for the series of computations during aircraft assembly simulation, allows to use more convenient formulations of the original problem (1) even though the preparation of data may take additional time. More specifically, the most efficient from a computational point of view are problems of smaller dimensions with simple constraints. Consider two different reformulations of problem (1), namely the dual problem and the relative problem.

The dual formulation of problem (1) is given by

$$\max_{\lambda} D(\lambda) = \max_{\lambda} -\tfrac{1}{2} \lambda^T Q \lambda + p^T \lambda + s,$$
$$\text{s.t. } \lambda \geq 0, \tag{3}$$

where $\lambda \in \mathbb{R}^m$ is the vector of Lagrange multipliers corresponding to contact forces in the area of possible contact, $Q = A^T K^{-1} A \in \mathbb{R}^{m \times m}$ is a symmetric positive definite dense matrix, $p = A^T K^{-1} f - g \in \mathbb{R}^m$ and $s = -\tfrac{1}{2} f^T K^{-1} f \in \mathbb{R}$.

The relative formulation of problem (1) is a variant of primal formulation that is based on a vector of relative displacements defined as $u = A^T x \in \mathbb{R}^m$:

$$\min_{u} R(u) = \min_{u} \tfrac{1}{2} u^T \tilde{K} u - \tilde{f}^T u,$$
$$\text{s.t. } u - g \leq 0, \tag{4}$$

38 M. Stefanova and S. Baklanov

where $\tilde{K} = (A^T K^{-1} A)^{-1} \in \mathbb{R}^{m \times m}$ is a symmetric positive-definite dense matrix and $\tilde{f} = \tilde{K} A^T K^{-1} f \in \mathbb{R}^m$.

The Theorem 1 shows the relationship between the primal (1), the dual (3), and the relative (4) formulations of the contact problem. All three formulations are equivalent, moreover, it follows from the theorem that the relative formulation (4) is a form of the primal problem, and the dual to the relative problem (4) is the dual to the primal problem (1). The relationship between different formulations, asserted by the Theorem 1, is established in [1] and [12].

Theorem 1. *The formulations of problem (1), (3) and (4) are equivalent, i.e. the vector x^* is the optimal solution to problem (1) if and only if there exists λ^* the optimal solution to problem (3) and there exists u^* the optimal solution to problem (4). Moreover, $P(x^*) = D(\lambda^*) = R(u^*)$, and the solutions of the problems are related by the relations $x^* = K^{-1}(f - A\lambda^*)$, $\lambda^* = \tilde{f} - \tilde{K}u^*$, $u^* = A^T x^*$.*

The advantage of the formulations (3) and (4) is the reduction in dimension from n unknowns to m. For example for two part joints, the unknowns and constraints number are related by $n = 2m$. The number of constraints m remains unchanged, but the linear constraints are replaced by the upper or lower bounds, which are less expensive for the used optimization methods [12]. At the same time, in the relative problem, the diagonal dominance of the Hessian matrix improves compared to the dual problem.

Preparation of matrices Q, $\tilde{K}$ and vectors p, $\tilde{f}$ requires additional computational costs. But since it is required to solve a set of contact problems (2) for the same matrices K and A, then the necessary matrices can be calculated once at a preliminary stage before solving QP problem. The construction of the relative problem turns out to be the most laborious. This is due to the fact that the ill-conditioned stiffness matrix K leads to a rapid accumulation of rounding errors when the matrix $A^T K^{-1} A$ is inverted without using error-free operations. As a result, inversion errors do not allow one to obtain a solution to the contact problem with the required accuracy.

3   Accurate Matrix Transformations

The current section briefly discusses methods for accurate inverse procedures for ill-conditioned matrices that are based on special algorithms for summation and dot product of floating point numbers that satisfy IEEE 754 format [7, 11]. The idea of error-free inversion methods is based on iterative reduction of inversion errors and use of additional memory to store the results of intermediate computations. All this significantly increases the time for preparing matrices for problem (4). The use of such laborious and expensive computations is justified due to the need to solve a large number of contact problems that differ only in assembly parameters, i.e. the vector of loads from fastening elements f and vector of initial gap g.

Addition and multiplication for floating point numbers are denoted as $fl(a + b)$ and $fl(a \cdot b)$, where $fl(\cdot)$ is the result of a floating point computation, where all operations inside the parentheses are executed in working (double) precision. Any sum of two floating point numbers a and b can be exactly represented using two floating point numbers c and d so that $a + b = c + d$, where $c = fl(a + b)$ and d is the exact addition error. Similarly, the multiplication of two floating point numbers a and b can be represented as $a \cdot b = c + d$, where $c = fl(a \cdot b)$ and d is multiplication error. Accounting for the exact errors d in matrix operations makes it possible to reduce computational errors. The corresponding methods [7] for computing c and error d are presented in Algorithms 1, 2, 3. The Algorithm 2 has two variants of realisation. In implementation for assembly modelling, the variant without using FMA gives a faster result.

Algorithm 1: TwoSum (Error-free transformation of the sum of two floating point numbers)

Data: Two floating point numbers a, b
Result: Two floating point numbers c, d, where $c = fl(a + b)$ and d is
        summation error
$c = fl(a + b)$;
$z = fl(c - a)$;
$d = fl(a - (c - z) + (b - z))$;

Algorithm 2: TwoMult (Error-free transformation of the product of two floating point numbers)

Data: Two floating point numbers a, b
Result: Two floating point numbers c, d, where $c = fl(a \cdot b)$ and d is
        multiplication error
$c = fl(a \cdot b)$;
$[a_1, a_2] = Split(a)$ (*);
$[b_1, b_2] = Split(b)$ (*);
$d = fl(a_2 \cdot b_2 - (((c - a_1 \cdot b_1) - a_2 \cdot b_1) - a_1 \cdot b_2))$ (*);
(lines marked with (*) can be replaced with $d = FMA(a, b, -c)$ for better performance, where $FMA()$ is the rounded-to-nearest exact result $a \cdot b - c$ (a build-in function for many processors));

Based on Algorithms 1 and 2, it is possible to perform error-free addition of several numbers, as well as the error-free dot product of two vectors [7] (Algorithms 4, 5, 6). The dot product and the summation Algorithms 5 and 4 have two additional inputs: k-fold precision and storage precision l. The larger k, the more accurate the result of the dot product, but at the same time, the more computational time is required. The larger l, the more accurately the result is

Algorithm 3: Split (Error-free splitting of a floating point number into two parts)

Data: The floating point number a
Result: Two floating point numbers x, y, where both x and y have at most $s - 1$ nonzero bits ($s = 27$ for double precision)
$factor = fl(2^s + 1);$
$c = fl(factor \cdot a);$
$x = fl(c - (c - a));$
$y = fl(a - x);$

stored, but the more memory is used and the longer it takes to process the result. Obviously, l have to be less than or equal to k.

The iterative algorithms for computing the inverse Cholesky decomposition are presented in [8] and matrix inversion in [11]. As part of this work, both of these methods were implemented to compute the Hessian $(A^T K^{-1} A)^{-1}$ for the relative problem (4). However, the latter (Algorithm 7) turned out to be more accurate. The Algorithm 7 uses the Frobenius norm, $inv(\cdot)$ is matrix inversion

Algorithm 4: SumKL (Error-free vector summation)

Data: The vector $p = \{p_1, ..., p_N\}$, where p_i are floating point numbers, K-fold precision, and storage precision l
Result: The summation res computed in k-fold precision and rounded into l parts
for $i = 1$ to $k - l$ **do**
$\quad | \quad p = VecSum(p);$
end
for $k = 0$ to $l - 2$ **do**
$\quad | \quad \{p_1, ..., p_{N-k}\} = VecSum(\{p_1, ..., p_{N-k}\});$
$\quad | \quad res_{k+1} = p_{N-k};$
end
$res_l = fl\left(\sum_i^{N-l+1} p_i\right);$

algorithm that is based on LU decomposition of M, $fl_{k,l}$ means matrix multiplication using the error-free dot product (Algorithm 5) with k-fold computation precision and storage precision l, ϵ is the machine epsilon. The number of iterations can also be taken as a stopping criterion, which helps to avoid the last multiplication.

Algorithm 5: VecMult (Error-free vector dot product)

Data: The vectors $p = \{p_1, ..., p_N\}$ and $q = \{q_1, ..., q_N\}$, where p_i and q_i are
floating point numbers, k-fold precision, and storage precision l
Result: Error-free dot product res of vectors p and q
for $i = 1$ *to* N **do**
  | $[r_i, r_{N+i}] = TwoMult(p_i, q_i)$;
end
$res = SumKL(r, k, l)$;

Algorithm 6: VecSum (Error-free vector transformation for summation)

Data: The vector $p = \{p_1, ..., p_N\}$, where p_i are floating point numbers

Result: The vector $q = \{q_1, ..., q_N\}$, such that $\sum_{i=1}^{N} p_i = \sum_{j=1}^{N} q_j$, where

$$q_N = fl(\sum_{i=1}^{N} p_i)$$

$q_1 = p_1$;
for $i = 2$ *to* N **do**
  | $[q_i, q_{i-1}] = TwoSum(p_i, q_{i-1})$;
end

Algorithm 7: MatrInv (Error-free matrix inversion)

Data: The matrix M
Result: The k-fold precision inversion R_k of matrix M, stored as the sum of k
matrices
$R_0 = \frac{1}{||M||} \cdot I$;
$k = 0$;
repeat
  | $k = k + 1$;
  | $P_k = fl_{k,1}(R_{k-1} \cdot M)$;
  | $X_k = inv(P_k)$;
  | $R_k = fl_{k,k}(X_k \cdot R_{k-1})$;
until $||X|| \cdot ||P|| < \epsilon^{-1}/100$;

To check the effect of error-free methods on the accuracy of the solution of
the contact problem, the solution of the primal problem (1) x, obtained on the
basis of the solution of the relative problem (4) by the Theorem 1, was analyzed.
The accuracy δ of the solution of the QP problem is determined based on the
Karush-Kuhn-Tucker optimality conditions:

$$\delta = \max\{||Kx - f + A\lambda||_\infty, \max\{a_i^T x - g_i, 0\}, \max\{-\lambda_i, 0\}, |(a_i^T x - g_i)\lambda_i|\},$$
$$i = 1..m, A = \{a_1, a_2, ..., a_m\}.$$

$$(5)$$

The following steps have been taken to improve solution accuracy of a relative problem (4):

Step 1: The error-free inversion of the matrix K (two iterations of Algorithm 7);

Step 2: The error-free inversion of the matrix $Q = A^T K^{-1} A$ (two iterations of Algorithm 7);

Step 3: The error-free dot product for multiplication $Q^{-1} \cdot A^T K^{-1} f$ (Algorithm 5);

Step 4: The storage of the matrix Q^{-1} obtained by Algorithm 7 as the sum of two matrices Q_1^{-1} and Q_2^{-1} ($l = 2$) and its usage for the multiplication $Q_1^{-1} \cdot A^T K^{-1} f + Q_2^{-1} \cdot A^T K^{-1} f$;

Step 5: The error-free dot product for multiplication $Q^{-1} A^T K^{-1} \cdot f$ (Algorithm 5).

Table 1. Comparison of accuracy, time and memory usage when using different 'Steps' to compute data for relative problem (4).

Steps	Accuracy δ	Computation time, sec.			max RAM, Mb
		K^{-1}	Q^{-1}	Total	
Without any steps	$8.9 \cdot 10^{-5}$	0.021	0.033	0.054	10.1
Step 1	$5.9 \cdot 10^{-5}$	2.2	0.033	2.233	10.1
Step 2	$5.4 \cdot 10^{-6}$	0.021	2.1	2.121	20.1
Steps 1,2	$1.1 \cdot 10^{-5}$	2.2	2.1	4.3	20.1
Steps 2,3	$2.7 \cdot 10^{-6}$	0.021	2.1	2.121	20.1
Steps 2,4	$3.0 \cdot 10^{-6}$	0.021	2.1	2.121	20.1
Steps 2,3,4	$3.6 \cdot 10^{-8}$	0.021	2.1	2.121	20.1
Steps 2,3,4,5	$1.4 \cdot 10^{-8}$	0.021	2.1	2.121	20.1
Steps 1,2,3,4,5	$6.5 \cdot 10^{-8}$	2.2	2.1	4.3	20.1

Table 1 compares the efficiency of the above steps in terms of computation time, the maximum used RAM and the value of solution accuracy δ. The comparison was carried out for a problem with the number of unknowns $n = 900$, the number of constraints $m = 573$ and the condition number of the Hessian matrix for the primal problem (1) $cond(K) = 1.26 \cdot 10^7$. The multiplication time

(Steps 3–5) is small compared to the inversion time even though multiplication is error-free. Thus, it does not affect inversion time.

The Steps 2,3,4,5 allow to obtain sufficiently accurate Hessian for relative problem (4). These actions require additional computational time and memory. The error-free inversion of the matrices K and Q (Steps 1,2) alone does not significantly increase the accuracy of the solution of the contact problem. The error-free inversion of the matrix K (Step 1) is not used in the final algorithm for solving the contact problem, since it does not affect the accuracy. Total time for accurate preparation of relative problem data takes about 40 times longer than usual. For solving QP problems and for recalculating the variables x, u, λ into each other, it is needed to store matrices K, K^{-1}, Q, Q^{-1}, $A^T K^{-1}$. Matrix inversion normal and error-free also uses auxiliary memory. Error-free inversion requires additional 5 square matrices (see Algorithm 7), namely P, X, the matrix for LU inverse in $inv(\cdot)$, R_1 and R_2. Normal inversion requires one matrix for Cholesky decomposition and its inversion. Note that the matrix K is block diagonal and is stored in blocks. The max RAM values presented in Table 1 do not simply reflect the total memory for all matrices used, but take into account intermediate deallocating and secondary usage of memory.

4   Algorithm for Solving the Contact Problem Using Relative Formulation

An algorithm that improves the accuracy of contact problem solution is presented below (Algorithm 8). The algorithm consists of two parts. The first one is related to the preparation of additional matrices necessary for the further solution of the QP problem, including error-free transformations needed for the relative formulation (6) of the QP problem. This step, as a preprocessor step, may be time intensive and does not affect the time of the actual variation simulation analysis. The second part presents numerous simulations of the assembly, which determine the total time of the variational simulation. Each i-th simulation is independent of the others and can be executed sequentially or in parallel [16]. Solution of relative problem (6) is performed using specially adapted to the assembly problem optimization methods [12]. As shown in [12], the Goldfarb-Idnani active set method [9] and the Newton projection method [1] are the fastest of considered methods when speaking about assembly problems. The algorithms are implemented in C++ as a part of ASRP software complex [6], developed for assembly modelling of riveting process.

Algorithm 8: Algorithm for solving the contact problem using relative formulation

---

Data: The reduced stiffness matrix K, the matrix of constraints A, the set of initial gaps g_i, the set of fasterning patterns f_i, where $i = 1..p$, p is number of assembly simulations

Result: The set of displacements x_i and contact forces λ_i

if *matrices* K^{-1}, $\tilde{K} = (A^T K^{-1} A)^{-1}$ *have been computed earlier* **then**

   | Load K^{-1}, $\tilde{K}$;

else

   Compute K^{-1} using Cholesky decomposition;

   Compute $\tilde{K}$ using error-free operations for inversion of $A^T K^{-1} A$ (Steps 2,4);

   Write matrices to the binary file on the disk;

end

for *i=1..p* **do**

   Compute $\tilde{f}_i = Q^{-1} A^T K^{-1} f_i$ using error-free multiplication (Steps 3,5);

   Solve relative problem:

$$\min_{u_i} \tfrac{1}{2} u_i^T \tilde{K} u_i - \tilde{f}_i^{\,T} u_i, \qquad (6)$$
$$\text{s.t. } u_i - g_i \le 0;$$

   Compute contact forces $\lambda_i = \tilde{f}_i - \tilde{K} u_i$;

   Compute displacement $x_i = K^{-1}(f_i - A\lambda_i)$;

   Check accuracy of x_i and λ_i by (5) (optional);

end

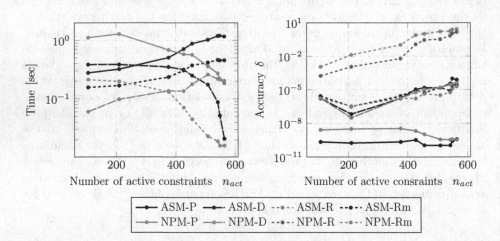

Fig. 2. Comparison of QP problem formulations for aircraft assembly problem by computation time (left) and accuracy (right). The lower wing-to-fuselage assembly model has $n = 900, m = 573, cond(K) = 1.26 \cdot 10^7$.

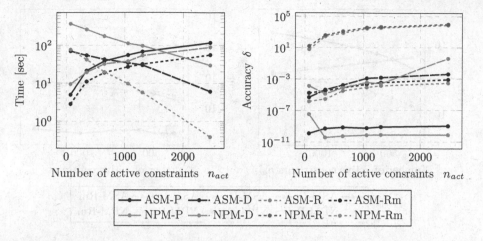

Fig. 3. Comparison of QP problem formulations for aircraft assembly problem by computation time (left) and accuracy (right). The lower wing-to-fuselage assembly model has $n = 4268, m = 2716, cond(K) = 1.2 \cdot 10^8$.

5   Comparison of Primal, Dual and Relative Formulations

By analogy with the Algorithm 8, the algorithms based on primal (1) and dual (3) formulation can be written, with the difference that error-free operations are not required. In this section, the solution of the contact problem obtained using different formulations of the QP problem, i.e. primal (1), dual (3) and relative (4) are compared. The active set method (ASM) [3,9,10] and the Newton projection method (NPM) [1,2] are used to solve the corresponding QP problem. The postfices '-P' , '-D', '-R' and '-Rm' stand for the primal problem (1), the dual problem (3), and the relative problem (4) without and with the use of 'Steps' (modified method, see Sect. 4). Algorithms based on different formulations of the QP problem are compared in terms of computation time and accuracy δ. Note that the accuracy, regardless of the used problem formulation, is always calculated relative to the original primal formulation. The stoping criteria of the methods are fixed and do not change for different QP problem formulation. The computation time includes only the time spent on one of the p assembly simulations and is shown in a logarithmic scale.

The comparison results are shown in the Figs. 2, 3, 4 and 5. The models of the most typical assembly joints are considered: upper and lower wing-to-fuselage model and fuselage section model. The figures show the computational time and accuracy depending on the number of active constraints n_{act} at the optimum. Note that different points corresponding to different n_{act} values refers to different fastening patterns f_i and initial gaps g_i for some representative simulations from the set of assembly problems (2). The highest accuracy is given by primal formulation and the accuracy value is several orders of magnitude higher compared to other formulations. The accuracy of the solutions obtained using the

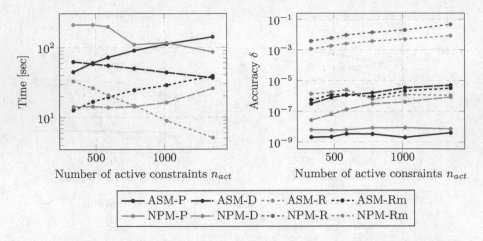

Fig. 4. Comparison of QP problem formulations for aircraft assembly problem by computation time (left) and accuracy (right). The upper wing-to-fuselage assembly model has $n = 6112, m = 3056, cond(K) = 1.1 \cdot 10^7$.

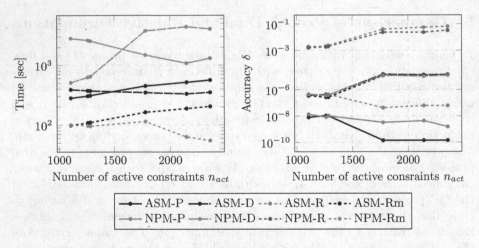

Fig. 5. Comparison of QP problem formulations for aircraft assembly problem by computation time (left) and accuracy (right). The fuselage section assembly model has $n = 11134, m = 6493, cond(K) = 1.12 \cdot 10^8$.

relative (using error-free operations) and dual formulations is close, while the computation time is much less compared to the primal formulation. Note, that for assembly problems accuracy higher than 10^{-4} is considered sufficient. The relative formulation in most simulations gives the fastest results.

6 Conclusion

The paper presents an modification of algorithm for solving the QP problem arising in assembly modeling problems. The algorithm is based on a relative reformulation of the contact problem. By the use of error-free operations at the stage of data preprocessing for relative QP problem, the accuracy of the related approach was increased. Importantly for this application field, the proposed approach allows to solve numerous number of QP problems quickly with a sufficient level of accuracy.

The development of an approach based on relative QP problem, together with the adaptation of optimization methods to the features of the assembly problem, helps to reduce the computation time of variation simulation analysis. Reducing the time, in turn, makes it possible to improve the assembly simulation by taking into account various additional physical effects, as well as to analyze and to optimize the assembly process.

Acknowledgements. The research was supported by Russian Science Foundation (project No. 22-19-00062, https://rscf.ru/en/project/22-19-00062/).

References

1. Baklanov, S., Stefanova, M., Lupuleac, S.: Newton projection method as applied to assembly simulation. Optim. Methods Softw. (2020)
2. Bertsekas, D.P.: Projected Newton methods for optimization problems with simple constraints. SIAM J. Control Optim. **20**(2), 221–246 (1982)
3. Goldfarb, D., Idnani, A.: A numerically stable dual method for solving strictly quadratic programs. Math. Program. **27**(1), 1–33 (1983)
4. Lorin, S., Lindau, B., Lindkvist, L., Söderberg, R.: Efficient compliant variation simulation of spot-welded assemblies. ASME J. Comput. Inf. Sci. Eng. **19**(1), 011007 (2019)
5. Lupuleac, S., Kovtun, M., Rodionova, O., Marguet, B.: Assembly simulation of riveting process. SAE Int. J. Aerosp. **2**, 193–198 (2010)
6. Lupuleac, S., et al.: Software complex for simulation of riveting process: Concept and applications. SAE Technical Paper, 2016-01-2090 (2016)
7. Ogita, T., Rump, S., Oishi, S.: Accurate sum and dot product. SIAM J. Sci. Comput. (SISC) **26**, 1955–1988 (2005)
8. Ogita, T., Oishi, S.: Accurate and robust inverse Cholesky factorization. Nonlinear theory and its applications. IEICE **3**(1), 103–111 (2012)
9. Petukhova, M.V., Lupuleac, S.V., Shinder, Y.K., Smirnov, A.B., Yakunin, S.A., Bretagnol, B.: Numerical approach for airframe assembly simulation. J. Math. Ind. **4**(8), 1–6 (2014)
10. Powell, M.J.D.: On the quadratic programming algorithm of Goldfarb and Idnani. In: Cottle, R.W. (ed.) Mathematical programming essays in honor of George B. Dantzig, Part II of the Springer Mathematical Programming Studies book series, vol. 25, pp. 46–61. Springer, Berlin, Heidelberg (1985). https://doi.org/10.1007/BFb0121074
11. Rump, S.M.: Inversion of extremely ill-conditioned matrices in floating-point. Japan J. Indust. Appl. Math. **26**, 249–277 (2009)

12. Stefanova, M., et al.: Convex optimization techniques in compliant assembly simulation. Optim. Eng. **21**(4), 1665–1690 (2020). https://doi.org/10.1007/s11081-020-09493-z
13. Wriggers, P.: Computational Contact Mechanics. Springer, Berlin (2006). https://doi.org/10.1007/978-3-540-32609-0
14. Yanagisawa, Y., Ogita, T., Oishi, S.: A modified algorithm for accurate inverse Cholesky factorization. Nonlinear Theory App. IEICE **5**, 35–46 (2014)
15. Yang, D., Qu, W., Ke, Y.: Evaluation of residual clearance after pre-joining and pre-joining scheme optimization in aircraft panel assembly. Assem. Autom. **36**(4), 376–387 (2016)
16. Zaitseva, N., Lupuleac, S., Petukhova, M., Churilova, M., Pogarskaia, T., Stefanova, M.: High performance computing for aircraft assembly optimization. In: 2018 Global Smart Industry Conference, pp. 1–6. IEEE, Chelyabinsk (2018)

Global Optimization

An Accelerated Algorithm for Finding Efficient Solutions in Multiobjective Problems with Black-Box Multiextremal Criteria

Konstantin Barkalov[✉][iD], Vladimir Grishagin[iD], and Evgeny Kozinov[iD]

Lobachevsky State University of Nizhni Novgorod, Nizhni Novgorod, Russia
{konstantin.barkalov,evgeny.kozinov}@itmm.unn.ru, vagris@unn.ru

Abstract. The paper is devoted to consideration of multiextremal multicriteria black-box problems with time-consuming criteria and proposes a new computational algorithm for solving such problems. This algorithm is based on combination of convolution approach, dimensionality reduction, and information-statistical global optimization algorithm. For approximating the Pareto set a family of maximum convolutions is considered generating problems of scalar multiextremal optimization. The latters are reduced by means of Peano mappings to the equivalent problems of univariate optimization that are solved by an accelerated information-statistical global optimization method. This method applies the technique of local refinements and uses in the course of current univariate optimization the information about criteria that was obtained when solving completed optimization problems. These features allow one to accelerate the search of global solution. The efficiency of the proposed algorithm is experimentally estimated on representative classes of multidimensional multiextremal test problems. Comparison with 5 nature-inspired particle swarm and genetic algorithms on the base of the hypervolume index reflecting quality of Pareto set evaluation demonstrates qualitative results of the proposed method.

Keywords: Multiobjective optimization · Global optimization · Dimensionality reduction · Search information · Nature-inspired metaheuristics · Numerical comparison

1 Introduction

In the field of mathematical modeling the intelligent decision-making processes, the problems of multicriteria optimization (MCO) present a wide class of models that are rich and interesting from a theoretical point of view and very important in real applications [16,18]. Multicriteria problems describe decision-making

The work was supported by the Ministry of Science and Higher Education of the Russian Federation (project No. 0729-2020-0055) and by the Research and Education Mathematical Center (project No. 075-02-2020-1483/1).

situations with multiple goals which are considered as functional criteria to be optimized.

In general case, there exist three key factors that determine the complexity of analyzing these models. The crucial feature corresponds to the typical case where criteria are contradictory: increasing one criterion leads to decreasing the other. Such inconsistent behavior generates the necessity of introducing a compromise in the joint analysis of criteria and consideration of a set of compromised (partial) solutions as a complete solution of a multiobjective problem (Pareto set), see, e.g., [5,8]. The second complexity factor is associated with the dimensionality or number of varied parameters of the problem because increasing the problem dimension leads to significant growth of computational cost.

Finally, the last factor is the multiextremality of criteria [24]. First of all, multiextremal criteria can generate complicated Pareto sets consisting, for example, of several disconnected parts. Moreover, the combination of such factors as multiextremality and greater dimension leads to significant computational complexity of the problem because the computational cost increases exponentially when dimension grows. From this point of view the multiextremal problems are the most complicated ones in multicriteria optimization and this MCO class is the subject of interest in the given paper.

The simplest way to build an approximation of the Pareto set consists in immersion in the feasible domain of problem's parameters a uniform grid (deterministic or random) and in selection of all non-dominated nodes of the grid. Unfortunately, this universal approach is too costly because it requires in the worst case pairwise comparison of all the grid nodes and does not take into account the properties of the problem for reducing the computational costs.

Another idea applies a parameterization scheme for the criteria of the problem and reduces a MCO problem to a family of single-criterion (scalar) optimization problems being problems of mathematical programming which can be solved by corresponding well-developed optimization methods [5,8]. In this scheme a non-negative coefficient (weight) is assigned to each criterion which reflects numerically an "importance" of the criterion and on the base of such weighted criteria a scalar function (convolution) is built and minimized. Under appropriate choice of convolution the solution of the built scalar optimization problem is a partial solution of the initial multicriteria problem. Changing weights we can obtain different Pareto points.

There exist various convolution functions but weighted sum function (linear convolution) and maximum from weighted criteria (maximum convolution) are most commonly used. The latter is more universal because it enables to build completely the Pareto set taking all the possible combinations of weights, whereas the linear convolution can lose Pareto points in non-convex case (in particular, if criteria are multiextremal).

For optimization of convolutions a wide spectrum of algorithms can be used in dependence on convolution properties generated by the features of criteria in the initial MCO problem. In particular, for the multiextremal case where it is required to find the global optimum the publications [2,9,11,13,36] describe applications of the global optimization algorithms to MCO problems in combination with the convolution approach. Among these methods of multiextremal optimization the efficient algorithms that utilize schemes of dimensionality reduction such as Peano mappings [31, 32] or nested optimization [14, 15, 32] are perspective for application in the frameworks of multiobjective optimization.

As for other methods of solving MCO problems it is necessary to pay attention to the important class of metaheuristic algorithms. Many of them are nature-inspired, i.e., are based on modeling physical processes (for example, simulated annealing [1, 17]) or mainly behavior of biological agents including evolutionary [4, 27] and as their part genetic algorithms [6, 28, 38] and particle swarm methods modeling the swarming intelligence of bees, fireflies, birds, ants, etc. [7, 21, 22].

Consideration of different approaches to solving MCO problems and a relevant references can be found, for example, in monographs and papers [8, 20, 23, 24, 35]. In this paper we consider multiextremal black-box MCO problems and propose a new scheme of solving these problems. This scheme is based on maximum convolution, dimensionality reduction by means of Peano curves, modified information-statistical global optimization algorithm used for the first time in MCO problems. Moreover, in the framework of the scheme the information already obtained in the course of solving scalar subproblems of convolution optimization can be utilized for solving a new convolution subproblem. Briefly, the way is as follows. Before beginning the optimization of a new convolution we have at our disposal values of criteria computed earlier and now can recalculate them to values of the current convolution taking into account these values as initial information for optimization algorithm. This procedure accelerates significantly the optimization process.

The quality of the proposed MCO method is experimentally studied on several multidimensional multiextremal test problems. Comparison with known evolutionary algorithms (2 particle swarm methods and 3 genetic algorithms have been taken) on the base of the hypervolume index [9, 36] reflecting quality of Pareto set estimation demonstrates qualitative results of the proposed method.

The rest of the paper is organized in the following manner. Section 2 considers the statement of the MCO problem to be studied. Section 3 is devoted to the description of the global search algorithm applied for scalar optimization of convolutions and the scheme of joint analyzing the family of ones. Section 4 contains results of computational experiments. The last Section concludes the paper.

2   MCO Problem Statement

The model of decision making to be considered hereinafter as the MCO problem consists in the following.

Let $w_i(x), 1 \leq i \leq k$, be real-valued functions depending on vector of arguments $x = (x_1, ..., x_N)$ and defined over the hyperparallelepiped

$$Q = \left\{ x \in R^N : a_j \leq x_j \leq b_j, 1 \leq j \leq N \right\} \tag{1}$$

in N-dimensional Euclidean space R^N. These functions reflect objectives of decision making and it is considered that the less the function value, the better the result of achieving this objective.

The problem of multicriteria optimization is formulated as minimizing over the domain (1) the vector function

$$W(x) = (w_1(x), ..., w_k(x)).\tag{2}$$

This problem can be symbolically written as

$$W(x) \to \min, \ x \in Q.\tag{3}$$

In terminology of decision making the vector function $W(x)$ is called *vector criterion* of the problem, functions $w_i(x), 1 \le i \le k$, are *partial criteria* or *objective functions*, Q is the domain of feasible decisions or just the *feasible domain* and points $x \in Q$ are *feasible decisions*.

As a rule, each criterion attains its minimum at a point different from minimum points of other criteria, i.e., it is impossible to find out the decision x^* providing in the region Q minimum values for all the partial criteria simultaneously. This situation generates the contradictoriness of criteria, when decreasing one criterion leads to the growth of the other, and complicates the notion of optimal decision (solution to the problem (3)).

Traditionally, optimal decisions in the problem (3) are defined on the base of Pareto optimality concept. To complete this item, let us give the known classical definitions concerning this concept.

Definition 1. *Let $x, z \in Q$. Vector x is said to dominate vector z $(x \succ z)$ if $w_i(x) \le w_i(z)$, $1 \le i \le k$, and there exists a number p, $1 \le p \le k$, such that $w_p(x) < w_p(z)$.*

Definition 2. *A decision vector $x^* \in Q$ is a Pareto optimal point if in the domain Q there is no vector $z \ne x^*$ dominating x^*.*

Definition 3. *The set of all Pareto optimal points (Pareto set P) is the Pareto optimal solution to the MCO problem (3).*

Definition 4. *The set $F = W(P) = \{W(x) : x \in P\}$ is called the Pareto front of the problem (3).*

As mentioned in Introduction, one way of building the Pareto set consists in reducing the multiobjective problem to a family of single-criteria, or scalar optimization problems and solving them by methods of mathematical programming. The global minimum point of a problem from this family will be a Pareto optimal point under corresponding assumptions. This approach can be realized by means of applying a convolution of the vector criterion (2), for example, the maximum convolution

$$\Gamma(\lambda, x) = \max_{1 \le i \le k} \lambda_i w_i(x),\tag{4}$$

where coefficients λ_i, $1 \leq i \leq k$, (weights of criteria) satisfy the conditions

$$\lambda_i \geq 0, \ 1 \leq i \leq k, \ \sum_{i=1}^{k} \lambda_i = 1. \tag{5}$$

If

$$w_i(x) > 0, \ 1 \leq i \leq k, \tag{6}$$

in the domain (1), then global solution to the problem

$$\Phi_\lambda(x) = \Gamma(\lambda, x) + \gamma \sum_{i=1}^{k} \lambda_i w_i(x) \to \min, \ x \in Q, \tag{7}$$

with a sufficiently small parameter $\gamma > 0$ is a Pareto optimal point of the initial MCO problem (3) [19,33]. The positiveness of criteria is not restrictive requirement because it is possible to transform easily the MCO problem with non-positive criteria to the form (6) without loss of Pareto solution.

In the case of multiextremal criteria the problem (7) is multiextremal as well and it is necessary to use global optimization algorithms for its solving. Many references to such the algorithms can be found in monographs [25, 26, 30–32, 34]. For solving multiextremal problems the information-statistical algorithms [31, 32] are among of the most efficient.

In the next section we propose a modified information-statistical algorithm with accelerated convergence and its application to searching for Pareto optimal solutions in multiextremal MCO problems.

3 Computational Scheme for Multiextremal Multiobjective Optimization

In this section we consider the MCO problem under additional assumptions according to which the criteria $w_i(x)$, $1 \leq i \leq k$, are black-box functions and satisfy in the domain Q the Lipschitz condition

$$|w_i(x) - w_i(y)| \leq L_i \|x - y\|, \ x, y, \in Q, \ 1 \leq i \leq k, \tag{8}$$

with corresponding Lipchitz constants $L_i > 0$ that are, as a rule, unknown. Here $\|*\|$ denotes the Euclidean norm in R^N.

The Lipschitzian functions are, in general case, multiextremal. If we consider the convolution (4) aggregating multiextremal criteria it is multiextremal as well.

In the computational scheme proposed in the paper we consider a family S of scalar problems (7) that are solved jointly and take into account the information obtained in the course of already completed optimizations. For this purpose in the region of weight coefficients determined by the conditions (5) a uniform grid is built and the family S consists of the problems (7) with coefficients λ corresponding to the grid nodes.

For solving the problems of the family, a global optimization technique based on reducing the multidimensional problem (7) to an equivalent univariate subproblem [25, 26, 30–32, 34] in combination with one-dimensional information algorithm with accelerated convergence is proposed.

Let us consider this technique in more detail.

To simplify the description let us present the problems (7) in the following unified form

$$f(x) \to \min, \ x \in Q, \tag{9}$$

where

$$f(x) = \max_{1 \le i \le k} \lambda_i w_i(x) + \gamma \sum_{i=1}^{k} \lambda_i w_i(x)$$

and $f(x)$ meets the Lipschitz condition

$$|f(x) - f(y)| \le L \|x - y\|, \ x, y, \in Q, \tag{10}$$

with the constant $L > 0$.

It is known as a fundamental fact (see, e.g., [29]) that N-dimensional hyperparallelepiped (1) and the interval $[0, 1]$ of the real axis are the equipotent sets and the interval $[0, 1]$ can be mapped onto the parallelepiped (1) unambiguously and continuously. Such mappings are called *Peano-type curves* or *evolvents*.

Let $x(t)$, $t \in [0, 1]$, be a Peano-type curve, and the function $f(x)$ from (9) be continuous. As

$$Q = \{x(t) : t \in [0, 1]\},$$

$f(x)$ and $x(t)$ are continuous, then

$$\min_{x \in Q} f(x) = \min_{t \in [0,1]} f(x(t)),$$

i.e., solving the multidimensional problem (9) can be reduced to the minimization of the one-dimensional function $f(x(t))$.

However, in the case when the function $f(x)$ is Lipschitzian the function $f(x(t))$ will satisfy the Hölder condition

$$|f(x(t')) - f(x(t''))| \le H |t' - t''|^{1/N}, \ t', t'' \in [0, 1], \tag{11}$$

with a Hölder constant $H > 0$. And if the function $f(x)$ satisfies the Lipschitz condition (10) then $H = 2L\sqrt{N + 3}$ (see [31, 32]). As a consequence, minimization of the function $f(x(t))$ requires the use of special algorithms oriented at functions with property (11).

In the proposed scheme for solving reduced univariate minimization problem we apply modified information-statistical Algorithm [32] with Local Refinements (ALR) accelerating convergence to global optimum of functions that satisfy the Hölder condition.

Before description of this algorithm let us follow again the path of simplification and formulate the general one-dimensional optimization problem as

$$\varphi(t) \to \min, \ t \in [\alpha, \beta]. \tag{12}$$

In our consideration $\varphi(t) = f(x(t))$, $[\alpha, \beta] = [0, 1]$ and $\varphi(t)$ meets the Hölder condition with the constant H.

The description of ALR can be presented in the following manner.

Let in the optimization problem the term "trial" denote the computation of objective function value at a point of the feasible domain.

According to ARL applied to solving the problem (12) its first $l \geq 2$ trials are executed at arbitrary points $t^1, \ldots, t^l$ of the interval $[\alpha, \beta]$ including points α and β, and the function values $v^1, \ldots, v^l$ are evaluated at these points, i.e., $v^s = \varphi(t^s)$, $1 \leq j \leq l$.

In order to choose a point t^{n+1} of a new $(n+1)$-th trial after completing $n \geq l$ trials it is necessary to implement the following actions.

Step 1. The points $t^1, \ldots, t^n$ of completed trials are ordered in increasing order and renumbered by subscripts, i.e.,

$$t_1 = \alpha < t_1 < \cdots < t_{n-1} < t_n = \beta. \tag{13}$$

The values $v_s = \varphi(t_s)$ are juxtaposed to the points t_s, $1 \leq s \leq n$.

Step 2. A real values $R(q)$ are assigned to all the intervals (t_{q-1}, t_q), $2 \leq q \leq n$, formed by neighboring points from (13). $R(q)$ is called *the characteristic* of the interval (t_{q-1}, t_q) and is computed in accordance with the following rule.

If number of completed trials is not exactly divisible by T, where *global-to-local ratio* T is a predefined number, then

$$R(q) = B(q), \tag{14}$$

otherwise

$$R(q) = B(q) \left(\frac{\sqrt{(v_q - \varphi^*(n))(v_{q-1} - \varphi^*(n))}}{M} + (1,5)^{-\sigma} \right)^{-1}. \tag{15}$$

Here

$$B(q) = (t_s - t_{s-1})^{1/N} + \frac{(v_q - v_{q-1})^2}{m^2(t_s - t_{s-1})^{1/N}} - 2\frac{v_q + v_{q-1}}{m}, \tag{16}$$

$$\varphi^*(n) = \min_{1 \leq j \leq n} \varphi(t^j)$$

is minimal computed value of the objective function $\varphi(t)$ at trial points,

$$M = \max_{2 \leq s \leq n} \frac{|v_s - v_{s-1}|}{(t_s - t_{s-1})^{1/N}} \tag{17}$$

is underestimation of the Hölder constant obtained according to the trial results,

$$m = \begin{cases} rM, & M > 0, \\ 1, & M = 0, \end{cases} \tag{18}$$

is improved estimation of the Hölder constant due to method's parameter $r > 1$, σ is some integer parameter that influences the speed of convergence to the global minimum.

Step 3. Among all intervals, the interval with the highest characteristic is selected

$$R(p) = \max_{2 \leq s \leq n} R(s). \tag{19}$$

Step 4. The new $(n+1)$-th trial is performed at the point

$$t^{n+1} = \frac{t_p - t_{p-1}}{2} - sign(v_p - v_{p-1}) \left(\frac{|v_p - v_{p-1}|}{M} \right)^N \frac{1}{2r}, \tag{20}$$

the value $v^{n+1} = \varphi(t^{n+1})$ is computed and the iteration number n is increased by 1.

The method ALR described above is a modification of the basic information-statistical algorithm [31,32] for minimization functions satisfying the Hölder condition. The modification consists in introducing new characteristics (15) after initial stage of the search to improve the method's speedup on account of increasing values of characteristics for intervals containing current record (16). Actually, the second term in (15) will have the largest value equal to $(1.5)^\sigma$ only for those intervals for which the equality $v_q = \varphi^*(n)$ or $v_{q-1} = \varphi^*(n)$ holds, i.e., for the intervals with the boundary points one of which coincides with the current record $\varphi^*(n)$.

Like its prototype [31,32], ARL can apply as the stopping condition the inequality

$$(t_p - t_{p-1})^{1/N} < \varepsilon, \tag{21}$$

i.e., it completes the search when the length of the interval with maximal characteristic (19) becomes less than the predefined accuracy $\varepsilon > 0$.

The sufficient condition of convergence to global minima in the problem (12) with the objective function under Hölder property with constant H is

$$m > 2^{2-1/N} H. \tag{22}$$

If $\varphi(t) = f(x(t))$ where $x(t)$ is the Peano-type curve and $f(x)$ meets (10), then $H = 2L\sqrt{N+3}$ and the sufficient condition takes the form

$$m > 2^{3-1/N} L\sqrt{N+3}. \tag{23}$$

The complete convergence theory of the information-statistical algorithms (including proof of the statements (22) and (23)) is given in [31,32].

Finalizing the proposed general scheme let us describe a procedure of successive solving the problems of the family S which are scalar problems (7) for different weight vectors $\lambda = (\lambda_1, \ldots, \lambda_k)$ (5). This procedure will be called Reduced Algorithm with Local Refinements and Data Accumulation (RALR-DA).

Solving the first problem from S we reduce it to the univariate problem (12) that is solved by the univariate method ALR. This method carries out 2 first iterations at boundary points and the next trials are executed in accordance with the points 1–4 of the algorithm's description up to stopping condition to be met.

In the course of optimization the algorithm generates the trial points $t^1, ..., t^n$ which correspond to the points $x^1 = x(t^1), ..., x^n = x(t^n)$ in the domain Q . At these points the values $v^j = \Phi_\lambda(x^j)$, $1 \leq j \leq n$, of the function $\Phi_\lambda(x)$ from (7) are computed as values $\varphi(t^j)$ of the function $\varphi(t)$ minimized by ALR.

In turn, obtaining the values $\Phi_\lambda(x^j)$ is implemented through the computation of criteria values $a_s^j = w_s(x^j)$, $1 \leq s \leq k$, $1 \leq j \leq n$, that can be considered as components of the matrix $A(k \times n)$ containing a posteriori information about the problem.

This information can be used for decreasing the number of trials during solving the problem (7) with other weight vector. Indeed, if we take the weight vector $\mu = (\mu_1, ..., \mu_k) \neq \lambda$, it is easy on the base of matrix A to calculate values $\Phi_\mu(x^j)$, $1 \leq j \leq n$, at points $x^1 = x(t^1), ..., x^n = x(t^n)$, i.e. values $\varphi(t^1), ..., \varphi(t^n)$ of the function $\varphi(t)$ in the problem (12) being the reduced problem (7) with weight vector μ. But in this case in ALR we can take points $t^1, ..., t^n$ as points of initial trials in which values $\varphi(t^j) = \Phi_\mu(x^j)$, $1 \leq j \leq n$, have been already computed and continue optimization following the rules 1–4 of the algorithm. In the course of continuation, ALR will generate new points t^q in which the values $\Phi_\mu(x(t^j))$, and, therefore, values $w_s(x(t^j))$, $1 \leq s \leq k$, will be evaluated. The latters can be added to the matrix A and be taken into account in the same manner for solving other problems (7).

The proposed scheme of utilizing a posteriori information allows one to decrease significantly the total number of criteria computation when joint optimizing problems of the family S. On the whole, the given approach is useful if the time spent to calculate the criteria at one point is significantly higher than the time taken by the algorithm for implementation of the trial at that point, i.e., it is oriented at time consuming MCO problems.

The efficiency of the described general scheme is estimated in the next section in comparison with several evolutionary methods.

4   Computational Experiments

For efficiency assessment of the computational scheme described in the previous section the experiment with multiextremal multidimensional MCO benchmarks was conducted. These benchmarks were constructed on the base of multiextremal functions belonging to the class GKLS [10] widely used in testing global optimization methods, i.e., the partial criteria $w_i(x)$, $1 \leq i \leq k$, were generated as GKLS functions.

Along with the proposed technique these MCO benchmarks were solved by 5 metaheuristic methods. Two algorithms – SMPSO [22] and OMOPSO [7] – belong to the class of particle swarm methods and 3 others are genetic ones (NSGA-II [6], SPEA2 [38], IBEA [37]). Implementation of these methods was taken from the library jMetalPy [3]. It should be noted that metaheuristic algorithms do not use convolutions to assess the Pareto set.

For estimating the effectiveness of solving the test MCO problem the hypervolume (HV) index [9] was used as the quality criterion of the Pareto set approximations built by the methods. Moreover, besides comparing the efficiency of tested algorithms via their HVs, an "ideal" value of HV was computed by means of selection of non-dominated nodes of a dense uniform grid placed in the feasible domain.

In total, three groups of experiments have been performed. Each group contained a series of MCO problems of the same dimension and number of criteria. Since all GKLS functions have a global minimum value equal to -1, the partial criteria $w_i(x)$, $1 \leq i \leq k$, were increased by a positive constant to ensure the requirements (6). In all experiments the domain Q from (1) with $a_j = -1$, $b_j = 1$ was considered as the feasible one.

Each problem was solved by all the methods including the grid technique and corresponding HVs were determined. Metaheuristic methods because of their stochastic nature sometimes realize unsuccessful runs. In order to estimate better their efficiency each of these methods carried out 10 launches when solving a separate MCO problem and the average result was taken into account.

During all experiments the method ALR was used with parameters $r = 5.6$, $\sigma = 18$, $T = 3$. The use of $T = 3$ corresponds to alternation of one "local" iteration and two "global" iterations, that can be interpreted as a combination of local refinement and global updating of the current optimizer. Accuracy ε from the stopping condition (21) was 0.05 in bi-criteria problems and 0.01 in 4-criteria ones.

For correct comparison of the algorithms (except grid method, of course) their parameters influencing the stopping condition were chosen to provide approximately the same number of trials. After the completion of solving all the group problems, HVs of each method were averaged.

In the first experiment 100 two-dimensional bi-criteria problems have been taken ($N = 2$, $k = 2$). The contour plots of a convolution $\Phi_\lambda(x)$ from (7) demonstrating its multiextremal and non-smooth relief are presented in Fig. 1. For estimating the Pareto set by RALR-DA, in each MCO problem 50 convolutions (7) corresponding to 50 values λ uniformly distributed over interval $[0, 1]$ have been minimized.

The Pareto front in one of test problems is shown in Fig. 2 and Fig. 3. Here Fig. 2 corresponds to the Pareto front computed on the uniform grid with the step 0.05 in the search domain whereas Fig. 3 shows the Pareto front computed by RALR-DA. The averaged HV indices obtained while solving the two-dimensional bi-criteria problems are presented in Table 1 (series 1) where column N is the dimension of the MCO problems, column k corresponds to criteria numbers and column "Pareto" corresponds to the estimation obtained by the grid method.

The second experiment was aimed at solving bi-criteria 4-dimensional MCO problems ($N = 4$, $k = 2$). 10 test problems have been investigated and for each of them, 100 convolutions (7) with uniformly distributed weight coefficients λ were minimized by RALR-DA. Table 1 (series 2) contains the results achieved by the methods participated in the experiment.

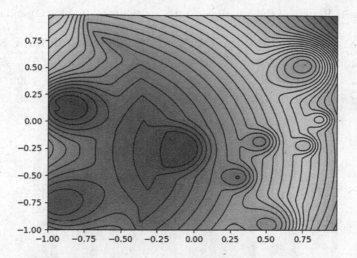

Fig. 1. Contour plots of 2-dimensional convolution.

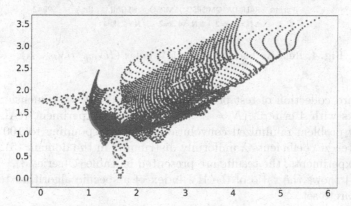

Fig. 2. Pareto front computed by grid technique.

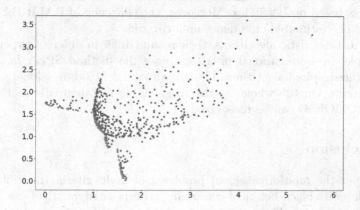

Fig. 3. Pareto front computed by RALR-DA.

Table 1. Results of the experiment with multidimensional MCO problems.

Series	N	k	Pareto	RALR-DA	OMOPSO	SMPSO	NSGAII	IBEA	SPEA2
1	2	2	6.62	6.44	5.93	5.88	5.71	5.65	5.61
2	4	2	5.04	4.71	4.45	4.37	4.39	4.39	4.53
3	2	4	23.89	23.52	22.87	22.53	19.82	20.68	22.80

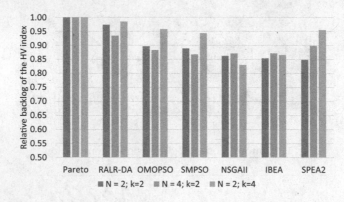

Fig. 4. Relative backlog of the HV index (HV_{alg}/HV_{Pareto}).

The third collection of test problems consisted of 10 two-dimensional MCO benchmarks with 4 criteria ($N = 2$, $k = 4$). In this experiment RALR-DA in each MCO problem minimized convolutions (7) corresponding to 100 different values of weight coefficients λ uniformly distributed in the domain (5). Like the previous experiments, the results are presented in Table 1 (series 3).

Figure 4 shows the ratio of the HV index of a specific algorithm to the HV index of Pareto set.

As it can be seen from the presented experiments, the proposed method demonstrates the best results among all methods involved in evaluation of their effectiveness based on HV index. Moreover, the efficiency of RALR-DA is close to the best one estimated on a dense uniform grid.

As for metaheuristic algorithms, their results differ in different experiments. For example, in 4-dimensional bi-criteria case the method SPEA is the best among nature-inspired algorithms, but it is the worst in 2-dimensional problems with 2 criteria. On the whole, as the most stable metaheuristic method, the algorithm OMOPSO can be recognized.

5   Conclusion

In the paper the multidimensional problems of multicriteria (multiobjective) optimization with black-box time-consuming criteria satisfying the Lipschitz condition are considered. These assumptions lead to significant complexity of the problems under investigation and require the development of special efficient numerical methods for their solving. For this goal in the paper a new algorithm is proposed and its computational scheme is described. The algorithm is based on reducing the MCO problem to a family of single-criterion (scalar) optimization problems generating Pareto-optimal solutions via building parameterized convolutions. In turn, each multidimensional scalar problem is reduced to a univariate one by means of Peano curves that map the multidimensional feasible domain into the real axis. Under Lipschitz condition for criteria the objective functions of the reduced one-dimensional problems satisfy the Hölder condition and for solving these problems it is necessary to apply specialized global optimization methods. As such method we propose a new algorithm that combines in its decision rule iteration with "local" and "global" goals. Global iterations provide the convergence to the global optimum and the local ones accelerate the search. Moreover, while solving the univariate problems generated by the family of convolutions the proposed algorithm is able to take into account the criteria values computed already in the course of solving the univariate problems completed earlier. This feature reduces significantly number of trials (calculations of criteria values): the later a problem is solved, the less new criteria computations are produced.

The quality of the proposed algorithm is studied in experiment on 3 representative sets of multiextremal MCO problems of different dimensions and criteria number in comparison with 5 particle swarm and genetic methods. As a criterion of methods' efficiency the hypervolume index estimating the quality of Pareto set approximation is taken. Among all competitors the proposed algorithm demonstrates the best results.

As the directions of further development of the current research it would be interesting to investigate the possibilities of applying other schemes of dimensionality reduction, for example, the nested optimization scheme [12,15], the diagonal [30] or simplicial [25] partition methods in the framework of the general approach used in the paper. Moreover, the prospective way is to develop parallel versions of the proposed general scheme oriented at powerful supercomputer systems.

References

1. Aarts, E., Korst, J., Michiels, W.: Simulated annealing. In: Floudas, C., Pardalos, P. (eds.) Encyclopedia of optimization, pp. 265–285. Springer, Boston (2014). https://doi.org/10.1007/978-1-4614-6940-7_10
2. Barkalov, K., Gergel, V., Grishagin, V., Kozinov, E.: An approach for simultaneous finding of multiple efficient decisions in multi-objective optimization problems. Lect. Notes Comput. Sci. **12755**, 127–143 (2021)
3. Benítez-Hidalgo, A., Nebro, A., García-Nieto, J., Oregi, I., Del Ser, J.: jMetalPy: a python framework for multi-objective optimization with metaheuristics. Swarm Evol. Comput.51(2019). https://doi.org/10.1016/j.swevo.2019.100598
4. Coello, C.A.C., Lamont, G.B., Van Veldhuizen, D.A.: Evolutionary Algorithms for Solving Multi-Objective Problems. Springer, New York (2007). https://doi.org/10.1007/978-0-387-36797-2

5. Collette, Y., Siarry, P.: Multiobjective Optimization: Principles and Case Studies (Decision Engineering). Springer, Berlin (2004). https://doi.org/10.1007/978-3-662-08883-8
6. Deb, K., Pratap, A., Agarwal, S., Meyarivan, T.: A fast and elitist multiobjective genetic algorithm: NSGA-II. IEEE Trans. Evol. Comput. **6**(2), 182–197 (2002). https://doi.org/10.1109/4235.996017
7. Durillo, J., García-Nieto, J., Nebro, A., Coello Coello, C., Luna, F., Alba, E.: Multi-objective particle swarm optimizers: An experimental comparison. Lect. Notes Comput. Sci. **5467**, 495–509 (2010). https://doi.org/10.1007/978-3-642-01020-0_39
8. Ehrgott, M.: Multicriteria Optimization. Springer, Berlin (2005). https://doi.org/10.1007/3-540-27659-9
9. Evtushenko, Y.G., Posypkin, M.A.: A deterministic algorithm for global multiobjective optimization. Optim. Methods Softw. **29**(5), 1005–1019 (2014). https://doi.org/10.1080/10556788.2013.854357
10. Gaviano, M., Kvasov, D.E., Lera, D., Sergeyev, Y.D.: Software for generation of classes of test functions with known local and global minima for global optimization. ACM Trans. Math. Softw. **29**(4), 469–480 (2003)
11. Gergel, V., Grishagin, V., Israfilov, R.: Adaptive dimensionality reduction in multiobjective optimization with multiextremal criteria. Lect. Notes Comput. Sci. **11331**, 129–140 (2019)
12. Gergel, V.P., Grishagin, V., Gergel, A.: Adaptive nested optimization scheme for multidimensional global search. J. Glob. Optim. **66**(1), 35–51 (2016)
13. Gergel, V.P., Kozinov, E.A.: Accelerating parallel multicriterial optimization methods based on intensive using of search information. Procedia Comput. Sci. **108**, 1463–1472 (2017). https://doi.org/10.1016/j.procs.2017.05.051
14. Grishagin, V., Israfilov, R.: Multidimensional constrained global optimization in domains with computable boundaries. In: CEUR Workshop Proceedings, vol. 1513, pp. 75–84 (2015)
15. Grishagin, V., Israfilov, R., Sergeyev, Y.D.: Convergence conditions and numerical comparison of global optimization methods based on dimensionality reduction schemes. Appl. Math. Comput. **318**, 270–280 (2018)
16. Hillermeier, C., Jahn, J.: Multiobjective optimization: survey of methods and industrial applications. Surv. Math. Ind. **11**, 1–42 (2005)
17. Locatelli, M.: Simulated annealing algorithms for continuous global optimization. In: Pardalos, P.M., Romeijn, H.E. (eds.) , Handbook of Global Optimization. Nonconvex Optimization and Its Applications, vol. 62, pp. 179–229. Springer, Boston (2002). https://doi.org/10.1007/978-1-4757-5362-2_6
18. Marler, R.T., Arora, J.S.: Multi-Objective Optimization: Concepts and Methods for Engineering. VDM Verlag (2009)
19. Marler, R., Arora, J.: Survey of multi-objective optimization methods for engineering. Struct. Multidiscip. Optim. **26**(6), 369–395 (2004). https://doi.org/10.1007/s00158-003-0368-6
20. Miettinen, K.: Nonlinear Multiobjective Optimization. Springer, Boston (1999). https://doi.org/10.1007/978-1-4615-5563-6
21. Mostaghim, S., Branke, J., Schmeck, H.: Multi-objective particle swarm optimization on computer grids. In: Proceedings of GECCO 2007: Genetic and Evolutionary Computation Conference, pp. 869–875 (2007). https://doi.org/10.1145/1276958.1277127

22. Nebro, A., Durillo, J., Nieto, G., Coello, C., Luna, F., Alba, E.: SMPSO: A new PSO-based metaheuristic for multi-objective optimization. In: Proceedings of MCDM 2009: IEEE Symposium on Computational Intelligence in Multi-Criteria Decision-Making, pp. 66–73 (2009). https://doi.org/10.1109/MCDM.2009.4938830

23. Nedjah, N., De Macedo Mourelle, L.: Evolutionary multi-objective optimisation: a survey. Int. J. Bio-Inspir. Comput. **7**(1), 1–25 (2015). https://doi.org/10.1504/IJBIC.2015.067991

24. Pardalos, P., Žilinskas, A., Žilinskas, J.: Non-Convex Multi-Objective Optimization. Springer, New York (2017). https://doi.org/10.1007/978-3-319-61007-8

25. Paulavičius, R., Žilinskas, J.: Simplicial Global Optimization. Springer, New York (2014). https://doi.org/10.1007/978-1-4614-9093-7

26. Pinter, J.D.: Global Optimization in Action (Continuous and Lipschitz Optimization: Algorithms, Implementations and Applications). Kluwer Academic Publishers, Dordrecht (1996)

27. Price, K.V., Storn, R.M., Lampinen, J.A.: Differential Evolution: A Practical Approach to Global Optimization. Natural Computing Series, Springer, New York (2005). https://doi.org/10.1007/3-540-31306-0

28. Ruiz, A.B., Saborido, R., Luque, M.: A preference-based evolutionary algorithm for multiobjective optimization: the weighting achievement scalarizing function genetic algorithm. J. Global Optim. **62**(1), 101–129 (2014). https://doi.org/10.1007/s10898-014-0214-y

29. Sagan, H.: Space-Filling Curves. Springer, New York (1994). https://doi.org/10.1007/978-0-387-39940-9_349

30. Sergeyev, Y.D., Kvasov, D.E.: Deterministic Global Optimization: An Introduction to the Diagonal Approach. Springer, New York (2017). https://doi.org/10.1007/978-1-4939-7199-2

31. Sergeyev, Y.D., Strongin, R.G., Lera, D.: Introduction to Global Optimization Exploiting Space-Filling Curves. Springer Briefs in Optimization, Springer, New York (2013). https://doi.org/10.1007/978-1-4939-7199-2

32. Strongin, R.G., Sergeyev, Y.D.: Global Optimization with Non-convex constraints Sequential and Parallel Algorithms. Kluwer Academic Publishers, Dordrecht (2000)

33. Wierzbicki, A.P.: The use of reference objectives in multiobjective optimization. In: Fandel, G., Gal, T. (eds.) Multiple Criteria Decision Making Theory and Application. pp. 468–486. Springer, Berlin (1980). https://doi.org/10.1007/978-3-642-48782-8_32

34. Zhigljavsky, A.A., Žilinskas, A.: Stochastic Global Optimization. Springer, New York (2008). https://doi.org/10.1007/978-0-387-74740-8

35. Zhou, A., Qu, B.Y., Li, H., Zhao, S.Z., Suganthan, P., Zhangd, Q.: Multiobjective evolutionary algorithms: a survey of the state of the art. Swarm Evol. Comput. **1**(1), 32–49 (2011). https://doi.org/10.1016/j.swevo.2011.03.001

36. Žilinskas, A., Žilinskas, J.: Adaptation of a one-step worst-case optimal univariate algorithm of bi-objective lipschitz optimization to multidimensional problems. Commun. Non-linear Sci. Numer. Simulat. **21**(1–3), 89–98 (2015)

37. Zitzler, E., Künzli, S.: Indicator-based selection in multiobjective search. Lect. Notes Comput. Sci. **3242**, 832–842 (2004). https://doi.org/10.1007/978-3-540-30217-9_84

38. Zitzler, E., Laumanns, M., Thiele, L.: SPEA2: improving the strength pareto evolutionary algorithm. TIK-Report 103 (2001). https://doi.org/10.3929/ethz-a-004284029

The Best Ellipsoidal Estimates of Invariant Sets for a Third-Order Switched Affine System

Alexander Pesterev$^{(\boxtimes)}$ (iD) and Yury Morozov (iD)

Institute of Control Sciences, Moscow 117997, Russia
alexanderpesterev.ap@gmail.com

Abstract. The switched affine system considered in the paper comes to existence when stabilizing the chain of three integrators subject to the additional condition of asymptotic tracking a desired trajectory. The target trajectory is defined implicitly as that of the second-order integrator stabilized by means of a feedback in the form of nested saturators. The control law for the third-order integrator is suggested that ensures the fulfillment of the additional condition, and the range of parameters of the proposed control guaranteeing global stability of the closed-loop system is determined. Moreover, the problem of constructing ellipsoidal estimates of invariant sets of the system is stated and solved. In particular, we consider the problem of finding the best (in a certain sense) ellipsoidal estimate that guarantees that the deviation of the system from the equilibrium point at any time does not exceed a prescribed value as long as the initial state vector falls into the ellipsoid.

Keywords: Invariant set · Ellipsoidal estimate · Chain of integrators · Affine switched system · Nested saturators

1 Introduction

The *continuous-time switched* (further, simply *switched*) *systems* form a particular class of the hybrid systems, the dynamical systems that exhibit characteristics of both continuous-time and discrete-time systems [1,2]. The switched affine system considered in the paper comes to existence when applying a piecewise continuous control of special form to a chain of three integrators. The goal of the control is to stabilize the system at the origin while making it follow a desired target trajectory in the course of the stabilization. The problem of stabilizing chains of integrators, as well as that of tracking a desired target trajectory, was widely discussed in the literature during last several decades (see, e.g., [3–6] and references therein). The interest to the problem is motivated by the fact that control scheme for chain of integrators can be easily extended to larger classes of systems (see, for example, [5]). Moreover, in many applications the nominal models have the form of chain of integrators, for instance, mechanical planar systems.

© The Author(s), under exclusive license to Springer Nature Switzerland AG 2022
N. Olenev et al. (Eds.): OPTIMA 2022, LNCS 13781, pp. 66–78, 2022.
https://doi.org/10.1007/978-3-031-22543-7_5

In this study, we sought for a control that globally stabilizes the third-order integrator under the additional condition of asymptotic tracking a desired trajectory when approaching the equilibrium state. The target trajectory is defined implicitly as that of a simpler second-order reference system stabilized by means of a feedback in the form of nested saturators, with the feedback coefficients being selected so that to ensure desired characteristics of the target trajectory. Another goal of this study was to develop an algorithm for constructing ellipsoidal estimates of the invariant sets of the closed-loop system. In particular, of interest are the best (in the sense of volume) estimates that guarantee that the deviation of the system from the equilibrium state in the course of stabilization does not exceed a prescribed value.

2   Problem Statement

2.1   Background

In [7–9], the problem of stabilizing a robot-wheel at a target point on a straight line subject to control and phase constraints was considered. By using the non-slipping condition and applying a continuous constrained control in the form of nested saturators, the problem was reduced to studying stability of the second-order integrator

$$\dot{w}_1 = w_2, \quad \dot{w}_2 = U_1(w_1, w_2), \tag{1}$$

subject to the control

$$U_1(w_1, w_2) = -k_4 \mathrm{sat}(k_3(w_2 + k_2 \mathrm{sat}(k_1 w_1))), \tag{2}$$

where $\mathrm{sat}(\cdot)$ is the nonsmooth saturation function: $\mathrm{sat}(w) = w$ for $|w| \leq 1$ and $\mathrm{sat}(w) = \mathrm{sign}(w)$ for $|w| > 1$. The coefficient k_4 is clearly the control resource U_{max}: $k_4 = U_{max}$. It is easy to prove (see, e.g., [7,8]) that $|w_2(t)| \leq k_2 \; \forall t > 0$ as long as $|w_2(0)| \leq k_2$; i.e., assuming that equation (1) governs motion of a mechanical system, k_2 is set equal to the maximum allowed velocity V_{max}: $k_2 = V_{max}$.

Near zero, system (1), (2) is linear,

$$\dot{w}_1 = w_2, \quad \dot{w}_2 = -k_1 k_2 k_3 k_4 w_1 - k_3 k_4 w_2,$$

and has a stable equilibrium at $w_1 = w_2 = 0$ for any $k_i > 0$, $i = 1, 2, 3, 4$. If $k_3 k_4 < 4k_1 k_2$, then the origin is a focus, otherwise, a node. The latter seems more practical since does not result in oscillations around the origin, and we further assume that the inequality $k_3 k_4 \geq 4k_1 k_2$ holds. System (1), (2) is known to be globally stable for any positive coefficients k_i [7,8].

Like in [8], to simplify calculations, we confine our consideration to the one-parameter family of the coefficients k_1 and k_3 by selecting them as

$$k_1 = \lambda/2k_2, \quad k_3 = 2\lambda/k_4, \quad \lambda > 0,$$

where $\lambda > 0$ is the desired exponential rate of the deviation decrease near the origin. Moreover, without loss of generality, the control resource and the maximum allowed velocity are set equal to ones, $k_2 = k_4 = 1$. With such a selection of the coefficients, system (1), (2) in the neighborhood of zero takes the form

$$\dot{w}_1 = w_2, \ \dot{w}_2 = -\lambda^2 w_1 - 2\lambda w_2, \tag{3}$$

with the origin being a stable degenerate node.

2.2   Stabilizing the Chain of Three Integrators

In this paper, we consider a more complicated problem of stabilizing a chain of three integrators:

$$\dot{x}_1 = x_2, \ \dot{x}_2 = x_3, \ \dot{x}_3 = U(x), \tag{4}$$

where $x \equiv [x_1, x_2, x_3]^{\mathrm{T}}$, by means of control $U(x)$ that ensures the fulfillment of the conditions

$$x_1(t) \to w_1(t), \ x_2(t) \to w_2(t), \tag{5}$$

where $(w_1(t), w_2(t))$ is the desired trajectory in the subspace of the coordinates x_1 and x_2, which is defined to be the trajectory of system (1), (2) with the initial conditions $w_1(0) = x_1(0), w_2(0) = x_2(0)$. In other words, we seek for a piecewise continuous control $U(x)$ that ensures asymptotic tracking of the trajectory of the second-order integrator stabilized by control (2) in the subspace of the first two coordinates. System (1), (2) will be further referred to as the *reference system* for system (4).

Comparing (1) and (4), one can see that the fulfillment of (5) is ensured if

$$x_3(t) \to U_1(x_1(t), x_2(t)), \ t \to \infty. \tag{6}$$

This, in turn, makes us suggest that the desired control can be sought in the form

$$U(x) = \dot{U}_1(x_1, x_2) + \gamma(U_1(x_1, x_2) - x_3), \tag{7}$$

where $\gamma > 0$ and $\dot{U}_1$ is the time derivative of function U_1 by virtue of system (1).

It seems practical to scale the coefficient γ in terms of λ, i.e., set $\gamma = \lambda\xi$, $\xi > 0$. Further, we will determine the range of parameters λ and ξ for which the closed-loop system (4), (7) is asymptotically stable, with conditions (6) being satisfied, and construct an invariant ellipsoidal estimate of the attraction domain.

2.3   Equivalent Switched Affine System

First, we show that the closed-loop system (4), (7) is classified among the class of linear switched affine systems. Consider the partitioning of the plane (x_1, x_2) into three sets (Fig. 1). In D_1, we include all points where both saturators are not saturated:

$$D_1 = \{(x_1, x_2) : |x_1| < 1/k_1, \ |x_2 + k_1 x_1| < 1/k_3\}$$

(the inclined strip bounded by the dashed lines in Fig. 1). The set D_2 consists of the points where the internal saturator reaches saturation, while the external one does not:

$$D_2 = \{(x_1, x_2) : |x_1| \geq 1/k_1, \ |x_2 + \text{sign}(x_1)| < 1/k_3\}$$

(the two disjoint horizontal strips shown in Fig. 1), and

$$D_3 = \{(x_1, x_2) : |x_2 + \text{sat}(k_1 x_1)| > 1/k_3\}$$

includes all points where the external saturator reaches saturation and, hence, $U_1 \equiv \pm 1$.

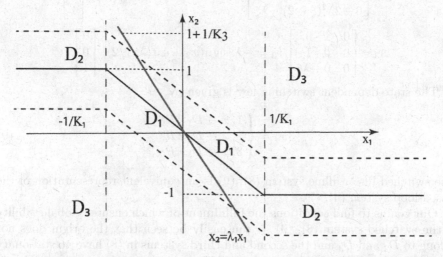

Fig. 1. Partitioning of the plane (x_1, x_2) into sets D_1, D_2, and D_3.

From (2), it is seen that function U_1 is piecewise linear:

$$U_1 = \begin{cases} -\lambda^2 x_1 - 2\lambda x_2, & (x_1, x_2) \in D_1, \\ -2\lambda(x_2 + \text{sign}(x_1)), & (x_1, x_2) \in D_2, \\ -\text{sign}(x_2 + \text{sat}(k_1 x_1)), & (x_1, x_2) \in D_3. \end{cases}$$

Taking the derivative of U_1 by virtue of system (1),

$$\dot{U}_1 = \frac{\partial U_1}{\partial x_1} x_2 + \frac{\partial U_1}{\partial x_2} U_1,$$

in the sets D_i, $i = 1, 2, 3$, we get

$$\dot{U}_1 = \begin{cases} 2\lambda^3 x_1 + 3\lambda^2 x_2, & (x_1, x_2) \in D_1, \\ 4\lambda^2(x_2 + \text{sign}(x_1)), & (x_1, x_2) \in D_2, \\ 0, & (x_1, x_2) \in D_3. \end{cases}$$

Substituting the expressions for U_1 and $\dot{U}_1$ into the equation for the closed-loop system (4), (7), we arrive at the affine switched system

$$\dot{x} = A_{i(x)}(\lambda,\xi)x + b_{i(x)}(\lambda,\xi), \ i(x) \in \{1,2,3\}, \tag{8}$$

where A_1, A_2, A_3 and b_1, b_2, b_3 are, respectively, constant matrices and vectors,

$$A_1 = \begin{bmatrix} 0 & 1 & 0 \\ 0 & 0 & 1 \\ -\lambda^3(\xi-2) & -\lambda^2(2\xi-3) & -\lambda\xi \end{bmatrix}, \ b_1 = 0,$$

$$A_2 = \begin{bmatrix} 0 & 1 & 0 \\ 0 & 0 & 1 \\ 0 & -2\lambda^2(\xi-2) & -\lambda\xi \end{bmatrix}, b_2 = -2\lambda^2(\xi-2)\mathrm{sign}(x_1)\begin{bmatrix} 0 \\ 0 \\ 1 \end{bmatrix},$$

$$A_3 = \begin{bmatrix} 0 & 1 & 0 \\ 0 & 0 & 1 \\ 0 & 0 & -\lambda\xi \end{bmatrix}, b_3 = -\lambda\xi\mathrm{sign}(x_2 + \mathrm{sat}(\lambda x_1/2))\begin{bmatrix} 0 \\ 0 \\ 1 \end{bmatrix}.$$

The state-dependent switching law is given by

$$i(x) = \begin{cases} 1, & x \in D_1 \times R, \\ 2, & x \in D_2 \times R, \\ 3, & x \in D_3 \times R. \end{cases} \tag{9}$$

The switched linear affine system (8), (9) is an equivalent representation of the closed-loop system (4), (7).

Our goal is to find conditions the fulfillment of which ensures global stability of the switched system (8), (9). It can easily be seen that the origin does not belong to D_2 and D_3 and the second and third systems in (8) have no stationary equilibria. Hence, the system can be stabilized at the origin only if matrix A_1 is Hurwitz, which yields necessary conditions for global stability of the system (which are also sufficient conditions for local stability). The following lemma is easily proved by calculating the characteristic polynomial of A_1 and applying the Routh–Hurwitz or Lienard–Chipart criterion [10].

Lemma 1. *System (8), (9) is locally asymptotically stable for any $\lambda > 0$ and $\xi > 2$.*

3   Global Stability

The proof of global stability is simplified if we replace x_3 by the new state variable δ_1:

$$\delta_1 = U_1(x_1, x_2) - x_3.$$

When deriving the equation in the new variable, we need the derivative of the function $U_1(x_1, x_2)$ by virtue of system (4). To distinguish it from the derivative

$\dot{U}_1$ by virtue of the second-order system (1), we denote the former as $\frac{d}{dt}U_1$. The relation between the two derivatives is easily found:

$$\frac{d}{dt}U_1 = \frac{\partial U_1}{\partial x_1}x_2 + \frac{\partial U_1}{\partial x_2}x_3 = \frac{\partial U_1}{\partial x_1}x_2 + \frac{\partial U_1}{\partial x_2}(U_1 - \delta_1) = \dot{U}_1 - \frac{\partial U_1}{\partial x_2}\delta_1.$$

Then, taking into account expression (7) for the control $U(x)$, it follows that

$$\dot{\delta}_1 = \frac{d}{dt}U_1 - \dot{x}_3 = -\frac{\partial U_1}{\partial x_2}\delta_1 - \lambda\xi\delta_1.$$

In the new variables, system (4), (7) takes the form

$$\dot{x}_1 = x_2, \quad \dot{x}_2 = U_1(x_1, x_2) - \delta_1, \tag{10}$$

$$\dot{\delta}_1 = -\left(\frac{\partial U_1}{\partial x_2} + \lambda\xi\right)\delta_1. \tag{11}$$

Equations (10) govern motion of the x-subsystem, which is a perturbed reference system (cf. Eq. (1)). It is easily seen that the components x_1, x_2 of the solution tend to the solution of the reference system (1) if and only if δ_1 tends to zero. With regard to the formulas

$$\frac{\partial U_1}{\partial x_2} = \begin{cases} -2\lambda, & (x_1, x_2) \in D_1, \\ -2\lambda, & (x_1, x_2) \in D_2, \\ 0, & (x_1, x_2) \in D_3, \end{cases}$$

equation (11) governing the δ-subsystem is the linear switched subsystem

$$\dot{\delta}_1 = -C_{i(x)}\delta_1 \tag{12}$$

with the state-dependent switching law (9), where $C_1 = C_2 = \lambda(\xi - 2)$ and $C_3 = \lambda\xi$. Since $C_1, C_2, C_3 > 0$ when $\xi > 2$, solution to Eq. (12) tends to zero for arbitrary switching laws. Hence, it follows that $\delta_1 = 0$ is the stable equilibrium of (12) for any state-dependent switching law either. Thus, we have proved that the necessary conditions of stability given by Lemma 1 are also sufficient ones for the global stability.

Theorem 1. *System (8), (9) is asymptotically stable in the large for any $\lambda > 0$ and $\xi > 2$.*

4 Construction of Ellipsoidal Estimates of the Invariant Sets

Asymptotic stability in the large is a very important characteristic of the system, but it does not forbid large deviations when converging to the equilibrium in the case of inappropriate initial conditions. Therefore, of interest are also invariant sets of the system. In particular, of great importance are sets inscribed in a strip $|x_1| \leq a$. If the initial state vector falls into such a set, then, for any

$t > 0$, the deviation of the system from the equilibrium point does not exceed a. Further in the paper, we consider construction of estimates of invariant sets of the system under study, which is based on results of absolute stability theory [11]. Particularly, we apply the technique similar to that developed in [12] (an example of its application to a two-dimensional system can be found in [13]). To take advantage of methods of absolute stability theory, we first rewrite the switched system (10), (11) in the form

$$\dot{x}_1 = x_2, \ \dot{x}_2 = -\Phi(\sigma, x), \ \dot{\delta}_1 = -C(x_1, x_2)\delta_1, \tag{13}$$

where $C(x_1, x_2) = C_1 = C_2$ when $(x_1, x_2) \in D_1$ or $(x_1, x_2) \in D_2$ and $C(x_1, x_2) = C_3$ when $(x_1, x_2) \in D_3$,

$$\Phi(\sigma, x) = \begin{cases} \sigma(x), & (x_1, x_2) \in D_1, \\ 2\lambda(x_2 + \operatorname{sign}(x_1)) + \delta_1, & (x_1, x_2) \in D_2, \\ \operatorname{sign}(x_2 + \operatorname{sat}(\lambda x_1/2)) + \delta_1, & (x_1, x_2) \in D_3, \end{cases} \tag{14}$$

$$\sigma(x) = \alpha^{\mathrm{T}} x, \ \alpha^{\mathrm{T}} = [\lambda^2, 2\lambda, 1] \tag{15}$$

(here, we use the same notation x for the new state vector, $x = [x_1, x_2, \delta_1]^{\mathrm{T}}$). It should be emphasized that, when $(x_1, x_2) \in D_1$, the stable linear system (13) in this notation takes the form

$$\dot{x}_1 = x_2, \ \dot{x}_2 = -\sigma(x), \ \dot{\delta}_1 = -C_1 \delta_1, \tag{16}$$

In what follows, we confine ourselves to constructing ellipsoidal estimates of the invariant sets

$$\Omega(P) = \{x : x^{\mathrm{T}} P x \le 1\}, \tag{17}$$

where P is a positive definite matrix of order three, under the additional condition constraining the deviation x_1:

$$\max_{x \in \Omega} |x_1| \le a. \tag{18}$$

Condition (18) is fulfilled by enforcing the following linear matrix inequality (LMI) [12]:

$$P \succ \begin{bmatrix} 1/a^2 & 0 & 0 \\ 0 & 0 & 0 \\ 0 & 0 & 0 \end{bmatrix}. \tag{19}$$

The remainder of the paper is devoted to solving the following

Problem: Find the best ellipsoidal estimate of the invariant set of system (13) inscribed in the strip $|x_1| \le a$.

Under the "best" ellipsoid, we further mean that the matrix P of which has minimal trace: $\operatorname{tr}(P) \to \min$.

Along with system (13), we consider the linear nonstationary system (*comparison system*)

$$\dot{x}_1 = x_2, \ \dot{x}_2 = -\beta(t)\sigma(x), \ \dot{\delta}_1 = -c(t)\delta_1, \tag{20}$$

where functions $\beta(t)$ and $c(t)$ satisfy the inequalities

$$0 < \beta_0 \leq \beta(t) \leq 1, \ C_1 \leq c(t) \leq C_3 \tag{21}$$

and conditions of existence and uniqueness of a continuous solution of the system. Let system (20) be absolutely stable, i.e., its zero solution is asymptotically stable for any functions $\beta(t)$ and $c(t)$ satisfying conditions (21).

Suppose now that, for any x, function $\Phi(\sigma, x)$ satisfies the "sector" condition

$$\beta_0 \sigma^2 \leq \Phi(\sigma, x)\sigma \leq \sigma^2. \tag{22}$$

The set of solutions of system (20) for all possible values of $\beta(t)$ and $c(t)$ satisfying conditions (21) is wider than that of the nonlinear system (13) with the function $\Phi(\sigma, x)$ satisfying (22) and $C(x_1, x_2)$ taking values $C_1 = C_2$ and C_3. Therefore, by requiring absolute stability of the zero solution of system (20), we ensure asymptotic stability of the zero solution of the nonlinear system (13), (22). Clearly, such an immersion of the system into a wider (in the sense of the set of solutions) class of systems yields only sufficient conditions.

Unfortunately, condition (22) is not fulfilled in the entire space R^3. Moreover, there does not exist a strip $|\sigma(x)| < \sigma'$, $\sigma' > 0$, where the sector condition holds for all x. Since, in [14], it is assumed that there exists a strip $-\sigma' < \sigma(x) < \sigma''$ in which the nonlinearity on the right-hand side of the system satisfies the sector condition, direct application of the method for constructing an attraction domain estimate proposed in [14] to the problem under study seems to be impossible.

However, as discussed in [12], the fulfilment of the sector condition (22) in the entire space R^3 is not actually required. For the invariant set of system (13), one can take an open *positive invariant set* [15] (further, simply invariant set) of the comparison system (20) if the sector condition (22) holds in this set. The invariant sets of system (20) can easily be found (see below), so that it remains to ensure the fulfillment of (22).

Thus, construction of an ellipsoidal estimate of the attraction domain for the nonlinear system (13) reduces to (i) finding a family of invariant ellipsoids of the absolutely stable linear nonstationary system (20) and (ii) selecting one of them with the matrix of minimal trace that belongs to a domain of system (13) in which the sector condition (22) holds. Clearly, the ellipsoid obtained in this way depends on the selected sector $[\beta_0, 1]$, where β_0 must be greater than the lower bound of the greatest sector of absolute stability of the system. Thus, we arrive at (iii) the problem of finding (estimating) the lowest $\beta_0 > 0$ for which the comparison system (20) is absolutely stable in the sector $[\beta_0, 1]$.

Solution of the first problem reduces to finding a quadratic Lyapunov function for the comparison system. Let us rewrite (20) in the matrix form as $\dot{x} = A_{[\beta,c]}(t)x$, where

$$A_{[\beta,c]}(t) = \begin{bmatrix} 0 & 1 & 0 \\ -\beta(t)\lambda^2 & -\beta(t)2\lambda & -\beta(t) \\ 0 & 0 & -c(t) \end{bmatrix}. \tag{23}$$

Let $A_{[1,C_1]}$, $A_{[1,C_3]}$, $A_{[\beta_0,C_1]}$, and $A_{[\beta_0,C_3]}$ denote the constant matrices obtained by the substitution of $\beta(t) \equiv 1, c(t) \equiv C_1$; $\beta(t) \equiv 1, c(t) \equiv C_3$; $\beta(t) \equiv \beta_0, c(t) \equiv C_1$; and $\beta(t) \equiv \beta_0, c(t) \equiv C_3$, respectively, into (23). In this notation, Eq. (16) is written as $\dot{x} = A_{[1,C_1]}x$. It is known [16,17] that the sufficient condition of absolute stability of system (20), (21) is the existence of a common Lyapunov function for the four linear systems $\dot{x} = A_{[1,C_1]}x$, $\dot{x} = A_{[1,C_2]}x$, $\dot{x} = A_{[\beta_0,C_1]}x$, and $\dot{x} = A_{[\beta_0,C_2]}x$. This assertion follows from the fact that, for any fixed t, matrix $A_{[\beta,c]}(t)$ can be represented as a convex linear combination of the four matrices.

In turn, a common quadratic Lyapunov function $\mathcal{L} = x^{\mathrm{T}}Px$ for these four systems exists if and only if the four LMIs

$$PA_{[1,C_1]} + A_{[1,C_1]}^{\mathrm{T}}P < 0, \ PA_{[\beta_0,C_1]} + A_{[\beta_0,C_1]}^{\mathrm{T}}P < 0,$$
$$PA_{[1,C_2]} + A_{[1,C_2]}^{\mathrm{T}}P < 0, \ PA_{[\beta_0,C_2]} + A_{[\beta_0,C_2]}^{\mathrm{T}}P < 0 \qquad (24)$$

have a nontrivial solution [16]. Hence, the sufficient condition of absolute stability of the linear nonstationary system (20), (21) is existence of a positive definite matrix P satisfying inequalities (24). Ellipsoid (17) the matrix of which satisfies LMIs (24) will be referred to as an *invariant ellipsoid of the comparison system* (20).

The second problem (ensuring the fulfillment of the sector condition in the ellipsoid) is solved by applying the technique developed in [12]. First, a domain with simple boundary (the boundary formed by first- and/or second-order surfaces) is identified where the sector inequalities (22) hold. Any ellipsoid with the matrix P defined from the LMI system (24) that belongs to such a domain is an invariant ellipsoid of system (13) if inequalities (22) hold. Inscribing the ellipsoid into a domain with simple boundary amounts to supplementing the LMI system (24) with additional m LMIs, where m is the number of surfaces forming the boundary [12].

It is easy to prove (the proof is omitted to save room) that the right-hand side of the sector condition (22) holds in the entire space R^3. Let us find the domain where the left-hand side of the sector condition holds; i.e.,

$$|\Phi(\sigma,x)| > \beta_0|\sigma|. \qquad (25)$$

Clearly, it trivially holds when $(x_1, x_2) \in D_1$ since $\Phi(\sigma,x) = \sigma$. Consider the case where $(x_1, x_2) \in D_3$. In this set, $\Phi = \delta_1 + \mathrm{sign}(\sigma_0)$. Introduce the notation

$$\sigma_0(x) = \lambda^2 x_1 + 2\lambda x_2,$$

so that $\sigma = \sigma_0 + \delta_1$.

Let $\sigma > 0$. Then, condition (25) yields:

$$\delta_1 + 1 > \beta_0\sigma_0 + \beta_0\delta_1,$$

or

$$\delta_1(1 - \beta_0) - \beta_0\lambda^2 x_1 - 2\beta_0\lambda x_2 > -1.$$

Similarly, for $\sigma < 0$, we obtain

$$\delta_1(1 - \beta_0) - \beta_0\lambda^2 x_1 - 2\beta_0\lambda x_2 < 1.$$

Combining both cases, we arrive at the conclusion that, for $(x_1, x_2) \in D_3$, condition (25) holds when

$$|g_1^{\mathrm{T}} x| \le 1, \ g_1^{\mathrm{T}} = [-\beta_0\lambda^2, -2\beta_0\lambda, 1 - \beta_0]. \tag{26}$$

i.e., in the strip confined by the planes $g_1^{\mathrm{T}} x = \pm 1$. The belonging of the ellipsoid to this strip is ensured by the LMI [12]

$$P > g_1 g_1^{\mathrm{T}}. \tag{27}$$

Now, let $(x_1, x_2) \in D_2$. In this case, $\Phi = 2\lambda(x_2 + \mathrm{sign}(x_1)) + \delta_1$. Let us denote

$$\sigma_\delta(x) = 2\lambda x_2 + \delta_1,$$

Let $x_1 > 0$. Then, condition (25) yields:

$$\sigma_\delta + 2\lambda > \beta_0\sigma_\delta + \beta_0\lambda^2 x_1,$$

or

$$(1 - \beta_0)\sigma_\delta - 2\beta_0\lambda^2 x_1 > -2\lambda.$$

Similarly, for $x_1 < 0$, we obtain

$$(1 - \beta_0)\sigma_\delta - 2\beta_0\lambda^2 x_1 < 2\lambda.$$

Combining both cases and substituting the formula for $\sigma_\delta(x)$, we arrive at the conclusion that, for $(x_1, x_2) \in D_2$, condition (25) holds when

$$|g_2^{\mathrm{T}} x| \le 1, \ g_2^{\mathrm{T}} = [-\beta_0\lambda/2, 1 - \beta_0, (1 - \beta_0)/2\lambda,]. \tag{28}$$

i.e., in the strip confined by the planes $g_2^{\mathrm{T}} x = \pm 1$. Like in the previous case, the LMI

$$P > g_2 g_2^{\mathrm{T}} \tag{29}$$

guarantees that the ellipsoid is inscribed into the strip between the planes $g_2^{\mathrm{T}} x = -1$ and $g_2^{\mathrm{T}} x = 1$.

Summing up the above discussion, we see that the desired ellipsoidal estimate of the invariant set is given by (17), where matrix P is the solution with the greatest trace of the set of seven LMIs: LMI (19), four LMIs (24), and LMIs (27) and (29). If $a \le 2/\lambda$, the strip $|x_1| \le a$ does not intersect with the set D_2. In this case, LMI (29) is not required, and matrix P is found by solving six LMIs.

The third task (finding the minimal β_0) is easily solved numerically. To this end, it is sufficient to numerically solve several times the LMI system (24) by incrementally decreasing the value of β_0 while the system has solutions and to take the largest value for which the system had solutions. Generally speaking, since the matrices involved in the LMI system depend on λ and ξ, the desired β_0 should also depend on these parameters. However, in our numerical experiments, the dependence on these parameters has not been observed: the minimal value of β_0 was approximately equal to 0.12 for all λ and ξ in the ranges if interest $0 < \lambda < 10$ and $2 < \xi < 10$, respectively. It is this value that was used in our numerical experiments below.

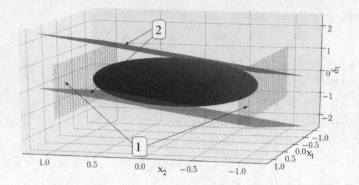

Fig. 2. Ellipsoid estimate $\Omega(P)$ of an invariant set inscribed in the strip bounded by the planes $|x_1| = \pm 0.95$ and the planes $g_1^T x = -1$ and $g_1^T x = 1$.

5   Numerical Example

To illustrate the discussion, we constructed the invariant ellipsoid inscribed in the strip $|x_1| \leq 0.95$ for the system with $\lambda = 2.2$ and $\xi = 3.2$. Since $a < 2/\lambda$, matrix P was sought by solving the system of LMIs (19), (24), and (27) with the help of the standard solver minimizing the trace of the matrix.

Figure 2 shows the ellipsoid found. The planes $x_1 = 0.95$ and $x_1 = -0.95$ are depicted in red color. The planes $g_1^T x = -1$ and $g_1^T x = 1$ are shown in green. As can be seen, the ellipsoid touches all four planes.

The cross section of the ellipsoid by the plane $z = 0$ is shown in Fig. 3. As can be seen, the "width" of the ellipsoid section in this plane is much greater than the width of the linearity set D_1 (boundaries marked by 3).

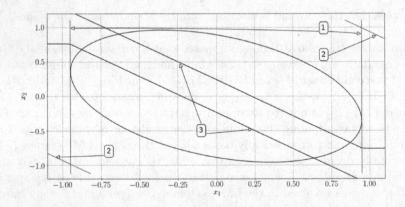

Fig. 3. The cross section of the ellipsoid $\Omega(P)$ by the plane $x_3 = 0$.

6 Conclusions

A piecewise continuous control law for stabilizing the chain of three integrators under the additional condition of asymptotic tracking a desired trajectory when approaching the equilibrium has been proposed. The target trajectory in the subspace of the first two coordinates is defined implicitly as that of the second-order integrator stabilized by means of a feedback in the form of nested saturators. The study of global stability of the closed-loop system was shown to reduce to studying stability of a switched affine system. The range of parameters of the proposed control guaranteeing global stability of the affine switched system has been determined. A method for constructing ellipsoidal estimates of invariant sets of the system based on results of absolute stability theory has been devised. The construction of an invariant ellipsoid reduces to solving a system of linear matrix inequalities.

References

1. Goebel, R., Sanfelice, R.G., Teel, A.R.: Hybrid Dynamical Systems: Modeling, Stability, and Robustness. Princeton University Press, Princeton (2012)
2. Lin, H., Antsaklis, P.J.: Stability and stabilizability of switched linear systems: a survey of recent results. IEEE Trans. Autom. Control **54**, 308–322 (2009)
3. Teel, A.R.: Global stabilization and restricted tracking for multiple integrators with bounded controls. Syst. Control Lett. **18**(3), 165–171 (1992)
4. Teel, A.R.: A nonlinear small gain theorem for the analysis of control systems with saturation. Trans. Autom. Contr. IEEE **41**, 1256–1270 (1996)
5. Polyakov, A., Efimov, D., Perruquetti, W.: Robust stabilization of MIMO systems in finite/fixed time. Int. J. Robust Nonlinear Control **26**(1), 69–90 (2016)
6. Kurzhanski, A.B., Varaiya, P.: Solution examples on ellipsoidal methods: computation in high dimensions. In: Dynamics and Control of Trajectory Tubes. SCFA, vol. 85, pp. 147–196. Springer, Cham (2014). https://doi.org/10.1007/978-3-319-10277-1_4
7. Pesterev, A.V., Morozov, Yu.V., Matrosov, I.V.: On optimal selection of coefficients of a controller in the point stabilization problem for a robot-wheel. In: Olenev, N., et al. (eds.) Communications in Computer and Information Science (CCIS), vol. 1340, pp. 236–249 (2020)
8. Pesterev, A., Morozov, Y.: Optimizing coefficients of a controller in the point stabilization problem for a robot-wheel. In: Olenev, N.N., Evtushenko, Y.G., Jaćimović, M., Khachay, M., Malkova, V. (eds.) OPTIMA 2021. LNCS, vol. 13078, pp. 191–202. Springer, Cham (2021). https://doi.org/10.1007/978-3-030-91059-4_14
9. Matrosov, I.V., Morozov, Yu.V. Pesterev, A.V.: Control of the robot-wheel with a pendulum. In: Proceedings of the 2020 International Conference "Stability and Oscillations of Nonlinear Control Systems" (Pyatnitskiy's Conference), pp. 1–4. IEEE Xplore, Piscataway, NJ (2020)
10. Gantmacher, F.: Matrix Theory. 5th edn. Fizmatlit, Moscow (2010)
11. Pyatnitskii, E.S.: Absolute stability of nonstationary nonlinear systems. Avtom. Telemekh. **1**, 5–15 (1970)
12. Pesterev, A.V.: Attraction domain estimate for single-input affine systems with constrained control. Autom Remote Control **78**, 581–594 (2017)

13. Pesterev, A.V.: Construction of the best ellipsoidal approximation of the attraction domain in stabilization problem for a wheeled robot. Autom. Remote. Control. **72**, 512–528 (2011)
14. Formal'skii, A.M.: Upravlyaemost' i ustoichivost' sistem s ogranichennymi resursami. Controllability and Stability of Systems with Constrained Resources. Nauka, Moscow (1974)
15. Blanchini, F., Miani, S.: Set-Theoretic Methods in Control. Birkhauser, Boston (2008)
16. Boyd, S., Ghaoui, L. E., Feron, E., Balakrishnan, V.: Linear Matrix Inequalities in System and Control Theory. SIAM, Philadelphia (1994)
17. Rapoport, L., Generalov, A.: On the stabilization of an inverted pendulum with a suspension point on the wheel. In: 16th International Conference on Stability and Oscillations of Nonlinear Control Systems (Pyatnitskiy's Conference), pp. 1–5. (2022)

Discrete and Combinatorial Optimization

Prize-Collecting Asymmetric Traveling Salesman Problem Admits Polynomial Time Approximation Within a Constant Ratio

Michael Khachay[1]([⊠]) [iD], Katherine Neznakhina[1,2] [iD], and Ksenia Rizhenko[1] [iD]

[1] Krasovsky Institute of Mathematics and Mechanics, Ekaterinburg, Russia
mkhachay@imm.uran.ru
[2] Ural Federal University, Ekaterinburg, Russia

Abstract. The Prize-Collecting Traveling Salesman Problem is an extension of the classic Traveling Salesman Problem, where each node of the given graph can be skipped for some known penalty. The goal is to construct a closed walk minimizing the total transportation costs and accumulated penalties. This problem has numerous applications in operations research, including sustainable production, supply chains, and drone routing. In this paper, we propose the first approximation algorithm with constant ratio for the asymmetric version of the problem on a complete weighted digraph, where the transportation costs fulfill the triangle inequality. Employing an arbitrary α-approximation algorithm for the Asymmetric Traveling Salesman Problem (ATSP) as a building block, our algorithm establishes an $(\alpha + 2)$-approximation for the Prize-Collecting Asymmetric Traveling Salesman Problem. In particular, using the seminal recent Swensson-Traub $(22 + \varepsilon)$-approximation algorithm for the ATSP, we obtain $(24 + \varepsilon)$-approximate solutions for our problem.

Keywords: Prize-Collecting Traveling Salesman Problem ·
Asymmetric TSP · Approximation with constant ratio

1 Introduction

The Prize-Collecting Traveling Salesman Problem (PCTSP) is one of the most well-studied problems in combinatorial optimization. Introduced by Egon Balas in his seminal paper [2], the problem has acquired a number of important applications in operations research including ones in metal production, drone routing [8], and ridesharing [19]. The problem has some connections to the classic Traveling Salesman Problem (TSP) [15] and Orienteering Problem (OP) [27].

Following [2], the problem has the following informal description. A salesperson is traversing some transportation network and collecting *a profit* for visiting

This research was carried out under the financial support of the Russian Science Foundation, grant no. 22-21-00672, https://rscf.ru/project/22-21-00672/.

'cities' (represented by the nodes of this network). However, for each unvisited node, he/she has to pay *a penalty*. Provided that a cost between any two cities u and v is known and non-negative, the salesperson wishes to construct a *closed walk* that gives at least a given value of profit and minimizes total transportation costs and penalties.

Related Work. As an extension of the classic TSP, the PCTSP remains to be strongly NP-hard even on the Euclidean plane [21] and hard to approximate in its general setting [24]. There are several main research directions in the field of algorithmic design for the PCTSP.

The first approach is concerned with exact Branch-and-Bound and Branch-and-Cut algorithms [4,9,10] and goes back to the fundamental facet-defining inequalities for the appropriate integer programs [2,12]. While being promising in terms of both hardware and software development, and having a significant impact on the combinatorial optimization, exact algorithms can be applied only to quite small instances of PCATSP.

The second approach is related to adaptations of different heuristics and metaheuristics including Tabu search [22], simulated annealing [23], VNS [17], genetic algorithms [11], and their combinations. These algorithms can be best performers quite so often and sometimes find the optimal (or really close sub-optimal) solutions in a few seconds for really large instances coming from the industry. The payoff, however, is the total lack of any theoretical guarantees. It creates additional labor costs related to numerical accuracy assessment and possible parameter tuning for any novel series of instances.

The third actively developed approach is based on the field of approximation algorithms with theoretical performance guarantees and approximation schemes. The first result here is 5/2-approximation algorithm for the metric PCTSP proposed by D. Bienstock, et al. [5]. Their algorithm is based on original rounding technique for the LP-relaxation of the problem and employs the classic Christofides-Serdyukov 3/2-approximation algorithm [7] for the metric TSP as a *black box*. Incorporation of the well-known primal-dual approach [14] made it possible to obtain several improvements of this result including $(1-2/3e^{-1/3})^{-1}$- and $(2-\varepsilon)$-approximation algorithms by [1,13], respectively.

Due to the fact that metric PCTSP is APX-hard, it appears that the best result that can be proved for arbitrary metrics (unless P=NP) is the approximation with constant ratio.

Nevertheless, for some special metrics there are indeed promising results, e.g. Polynomial-Time Approximation Schemes (PTAS) for the PCTSP on planar graphs [3] and the PCTSP formulated in metric spaces of an arbitrary fixed doubling dimension [6].

On the other hand, for the asymmetric version of the problem, we call it the Prize-Collecting Asymmetric Traveling Salesman Problem (PCATSP), approximation algorithms with theoretical performance guarantees still remain to be quite rare. To the best of our knowledge, the only result is $(1 + \lceil \log n \rceil)$-approximation algorithm [20]. In this paper, we do our best to bridge this gap.

Our Contribution. We propose the first approximation algorithm with constant ratio (with $(24 + \varepsilon)$ accuracy upper bound) for the PCATSP with triangle inequality. Our approach extends the rounding framework proposed in [5] and employs the recent Svensson-Traub $(22 + \varepsilon)$-approximation algorithm for the ATSP with triangle inequality as a building block.

In Sect. 2 we introduce definitions and notation and remind some known results the necessary for the subsequent constructions. In Sect. 3, we present a description of the PCATSP and the proposed MILP-model. Then, in Sect. 4, we discuss the proposed approximation algorithm, whose accuracy and time complexity bounds are proved in Sect. 5. Finally, in Sect. 6, we summarize our results and discuss some questions still remaining open.

2   Preliminaries

Our results are based on approximation algorithm for the Asymmetric Traveling Salesman Problem (ATSP) with triangle inequality. An instance of the ATSP is given by a complete edge-weighted digraph $G = (V, A, c)$, where, similarly to the PCATSP, the weighting function $c\colon A \to \mathbb{R}_+$ specifies transportation costs and satisfies the triangle inequality

$$c(u, v) + c(v, w) \geq c(u, w) \text{ for any } \{u, v, w\} \subset V. \tag{1}$$

The goal is to construct a minimum cost *tour* T (a closed walk visiting all the nodes of the graph G).

Also, it is convenient to consider an alternative but equivalent formulation of the ATSP, where it is required to find a (multi-)subset $T \subset A$ of the minimum cost

$$c(T) = \sum_{a \in T} c_a x_a. \tag{2}$$

In (2), variable x_a denotes multiplicity of the arc $a \in T$ and the directed multigraph (V, T) is Eulerian and strongly connected. In the following, we do not distinguish a tour T and the corresponding arc-multiplicity vector x.

Recall some necessary definitions and concepts. For arbitrary nonempty subsets $U_1, U_2 \subset V$, we use a standard notation

$$\delta(U_1, U_2) = \{(u, v)\colon u \in U_1, v \in U_2\}.$$

In particular, for $U_1 = U$ and $U_2 = V \setminus U$, we obtain *outgoing* and *incoming cuts*

$$\delta^+(U) = \delta(U, V \setminus U) = \{(u, v) \in A\colon u \in U, v \notin U\} \text{ and}$$
$$\delta^-(U) = \delta(V \setminus U, U) = \{(u, v) \in A\colon u \notin U, v \in U\},$$

respectively and the cut $\delta(U) = \delta^+(U) \cup \delta^-(U)$. If $U = \{v\}$, we use a simplified notation $\delta(v)$.

In the following, we need the classic Held-Karp MILP-model of the ATSP

$$\min \sum_{a \in A} c_a x_a \tag{3}$$

$$s.t. \quad x(\delta^+(v)) = x(\delta^-(v)) \quad (v \in V) \tag{4}$$

$$x(\delta(U)) \geq 2 \quad (\emptyset \neq U \subset V) \tag{5}$$

$$x_a \in \mathbb{Z}_+ \quad (a \in A). \tag{6}$$

Here, Eq. (4) ensures that each feasible solution corresponds to the Eulerian multi-graph and (5) are the well-known subtour elimination constraint. Let ATSP* and ATSP$_{LP}^*$ denote optimum values of problem (3)–(6) and its LP-relaxation, respectively.

Recently Svensson et al. [25] proved that the ATSP can be approximated in polynomial time within a constant ratio. Since then, several subsequent improvements of this breakthrough result were proposed. The state-of-the-art approximation algorithm for this problem was proposed by V. Traub and J. Vygen in [26]. We summarize their result in the following theorem.

Theorem 1. *For an arbitrary $\varepsilon > 0$, there exists a polynomial time algorithm for the ATSP with triangle inequality that finds a feasible solution R of cost*

$$\text{ATSP}^* \leq c(R) \leq (22 + \varepsilon)\,\text{ATSP}_{LP}^*.$$

In the sequel, we employ Theorem 1 to prove the similar approximation result for the PCATSP.

3   Problem Statement

We consider a simplified version of the Prize-Collecting Asymmetric Traveling Salesman Problem (PCATSP) introduced in [5] and also known as Profitable Tour Problem (PTP) [10].

An instance of the PCATSP is given by a node- and edge-weighted complete digraph $G = (V, A, c, \pi)$, where the weighting function $c\colon A \to \mathbb{R}_+$ specifies transportation costs and fulfills the triangle inequality (1) and the function $\pi\colon V \to \mathbb{R}_+$ defines penalties for skipping nodes of the graph G.

Unlike the classic ATSP, in PCATSP, any closed walk, including an *empty* one that skips all the nodes and a *degenerate* walk (a loop of zero transportation costs) skipping all the nodes except one, is feasible.

For an arbitrary walk W, its cost is defined as follows:

$$\text{cost}(W) = \begin{cases} \sum_{a \in W} c_a + \sum_{v \notin V(W)} \pi_v, & \text{if } W \text{ is a non-degenerate walk}, \\ \sum_{v \notin V(W)} \pi_v, & \text{otherwise}, \end{cases}$$

the subset $V(W) \subset V$ consists of the nodes visited by the walk W.

The goal is to find the cheapest feasible walk.

The main differences between the considered formulation and the original one given by Balas are as follows:

(i) we consider transportation costs subject to triangle inequality (1),
(ii) we assign zero profits for visiting the cities (nodes of the graph G),
(iii) knapsack constraint restricting minimum gathered profit is excluded.

By extending the known approach (see, e.g. [5,9]), we propose the following MILP-model for the PCATSP:

$$\min \sum_{a \in A} c_a x_a + \sum_{w \in V} \pi_w (1 - y_w) \tag{7}$$

$$s.t. \qquad x(\delta^+(w)) - x(\delta^-(w)) = 0 \ (w \in V), \tag{8}$$

$$x(\delta(S)) \geq 2y_v + 2y_w - 2 \ (S \subseteq V, v \in S, w \notin S) \tag{9}$$

$$x_a \in \mathbb{Z}_+, \ y_w \in \{0,1\}. \tag{10}$$

In the model,

- Boolean variable y_w indicates whether the node $w \in V$ is visited by the walk;
- the objective function (7) represents transportation costs and accumulated penalties for skipping the nodes;
- Equation (8) ensures that the multi-graph (V,T) is Eulerian and, together with (9), strongly connected.

As above, we introduce an LP-relaxation $\text{PCATSP}_{\text{LP}}$ of the problem PCATSP, where constraint (10) is replaced by $x_a \geq 0$ and $0 \leq y_w \leq 1$. Also, by PCATSP^*, $\text{PCATSP}_{\text{LP}}^*$ we denote optimum values of the problems PCATSP and $\text{PCATSP}_{\text{LP}}$, respectively.

4   Approximation Algorithm

In this section, we describe the proposed approximation Algorithm $\mathcal{A}$ for the PCATSP. Algorithm $\mathcal{A}$ has two external parameters

- a polynomial time approximation Algorithm $\mathcal{A}_0$ for the ATSP, which finds in time $\mathcal{T}(\mathcal{A}_0)$ an approximate solution of cost APP, such that

$$\text{ATSP}^* \leq \text{APP} \leq \alpha \, \text{ATSP}_{\text{LP}}^*, \tag{11}$$

for some $\alpha \geq 0$;
- a threshold τ specifying the minimum size of an auxiliary ATSP instance, for which we use Algorithm $\mathcal{A}_0$ for finding its approximate solution; all the instances of smaller sizes are solved to optimality in constant time by an arbitrary exact algorithm.

Algorithm $\mathcal{A}$

    Input: an instance of the PCATSP
    Parameters: an approx. algorithm $\mathcal{A}_0$ for the ATSP with triangle inequality,
                a threshold $\tau \geq 2$
    Output: an approx. solution $\hat{W}$ of the given instance
1: find an optimum solution $(\bar{x}, \bar{y})$ for the $\text{PCATSP}_{\text{LP}}$
2: consider an auxiliary ATSP instance $\mathcal{I}_\alpha$ specified by the subgraph $G\langle V_\alpha \rangle$ induced
   by the subset

$$V_\alpha = \left\{ w \in V : \bar{y}_w \geq \frac{\alpha + 1}{\alpha + 2} \right\}$$

3: output the walk $\hat{W}$ defined as follows:

$$\hat{W} = \begin{cases} \text{the approximate solution of } \mathcal{I}_\alpha \text{ provided by Algorithm } \mathcal{A}_0, \text{ if } |V_\alpha| \geq \tau, \\ \text{an arbitrary optimal solution of } \mathcal{I}_\alpha, \text{ if } 2 \leq |V_\alpha| < \tau, \\ \text{the degenerate walk (loop) } (v), \text{ if } V_\alpha = \{v\} \text{ for some } v \in V, \\ \text{the empty walk, otherwise.} \end{cases}$$

---

5   Approximation Ratio and Complexity Bounds

In this section, we show that Algorithm $\mathcal{A}$ has the similar approximation properties as Algorithm $\mathcal{A}_0$. In particular, if we take as Algorithm $\mathcal{A}_0$ the Svensson-Traub algorithm, then Algorithm $\mathcal{A}$, in polynomial time, will obtain a feasible solution of the problem PCATSP, whose cost does not exceed $(24 + \varepsilon) \, \text{PCATSP}_{\text{LP}}{}^*$.

Theorem 2. *Algorithm $\mathcal{A}$ finds an approximate solution $\hat{W}$ of the PCATSP, such that*

$$\text{PCATSP}^* \leq \text{cost}(\hat{W}) \leq (\alpha + 2) \, \text{PCATSP}_{\text{LP}}{}^*, \tag{12}$$

in time $\mathcal{T}(\mathcal{A}_0) + \mathcal{T}(\text{PCATSP}_{\text{LP}})$, where $\mathcal{T}(\text{PCATSP}_{\text{LP}})$ is time complexity of the linear program $\text{PCATSP}_{\text{LP}}$.

Proof. Since an upper bound for the time complexity of Algorithm $\mathcal{A}$ easily follows from its description, we proceed with approximation ratio bound (12). In the case $V_\alpha = \varnothing$, Algorithm $\mathcal{A}$ outputs the empty walk W_0 of $\text{cost}(W_0) = \sum_{v \in V} \pi_w$. Therefore,

$$\frac{\text{cost}(W_0)}{\text{PCATSP}_{\text{LP}}{}^*} = \frac{\sum\limits_{w \in V} \pi_w}{\sum\limits_{a \in A} c_a \bar{x}_a + \sum\limits_{w \in V} \pi_w (1 - \bar{y}_w)}$$

$$\leq \frac{\sum\limits_{w \in V} \pi_w}{\sum\limits_{a \in A} c_a \bar{x}_a + (\alpha + 2)^{-1} \sum\limits_{w \in V} \pi_w} \leq (\alpha + 2).$$

In the case $V_\alpha = \{v\}$ for an arbitrary $v \in V$, the desired upper bound can be obtained similarly.

Consider the common case, where $|V_\alpha| \geq \tau$. To proceed, transform the fractional solution $(\bar{x}, \bar{y})$ found at Step 1 of Algorithm $\mathcal{A}$ as follows:

$$\hat{x} = \left(\frac{\alpha+2}{\alpha}\right)\bar{x}; \quad \hat{y}_w = \begin{cases} 1, & \text{if} \quad \bar{y}_w \geq (\alpha+1)/(\alpha+2), \\ 0, & \text{otherwise} \end{cases} \quad (w \in V). \quad (13)$$

By construction, the set V_α defined at Step 2 of Algorithm $\mathcal{A}$ fulfills the equation

$$V_\alpha = \{w \in V : \hat{y}_w = 1\}.$$

Further, the walk $\hat{W}$ obtained at Step 3 is an approximate solution of the auxiliary instance $\mathcal{I}_\alpha$ of the ATSP specified by the subgraph $G\langle V_\alpha \rangle$ provided by Algorithm $\mathcal{A}_0$. By x' we denote the appropriate feasible solution of the MILP-model:

$$\min \sum c_a x_a \tag{14}$$
$$s.t. \quad x(\delta^+(w)) - x(\delta^-(w)) = 0 \quad (w \in V_\alpha), \tag{15}$$
$$x(\delta(U)) \geq 2 \quad (\emptyset \neq U \subset V_\alpha), \tag{16}$$
$$x_a \in \mathbb{Z}_+. \tag{17}$$

By (11), we have

$$\text{ATSP}^* \leq c(x') \leq \alpha \cdot \text{ATSP}_{\text{LP}}^*,$$

where ATSP^* is an optimum of the instance $\mathcal{I}_\alpha$ and $\text{ATSP}_{\text{LP}}^*$ is an optimum of its LP-relaxation $\mathcal{I}_{\text{LP}}$. Next, it is easy to verify that the problem $\mathcal{I}_{\text{LP}}$ is equivalent to the following linear program:

$$\min \sum c_a x_a \tag{18}$$
$$s.t. \quad x(\delta^+(w)) - x(\delta^-(w)) = 0 \quad (w \in V), \tag{19}$$
$$x(\delta(S)) \geq 2 \quad (S \subset V : V_\alpha \cap S \neq \emptyset, V_\alpha \setminus S \neq \emptyset), \tag{20}$$
$$x_a \geq 0 \tag{21}$$
$$x(\delta(w)) \begin{cases} \geq 2, & \text{if } w \in V_\alpha \\ = 0, & \text{otherwise} \end{cases} \quad (w \in V). \tag{22}$$

By evolving the idea of the proof Lemma 2.2 from [5] and adopting the seminal results obtained by Lovász [18] and Jackson [16] for connectivity of Eulerian graphs, we show that the optimum of problem (18)–(22) coincides with the optimum of problem (18)–(21). For the sake of brevity, we skip this part of the proof postponing it to the forthcoming full paper.

To complete the proof, we should show that the vector $\hat{x}$ is a feasible solution of problem (18)–(21). Indeed, Eqs. (19) and (21) easily follow from (8) and (10) respectively, since $\hat{x} = (\alpha + 1)/\alpha \cdot \bar{x}$. To prove that $\hat{x}$ fulfills inequality (20) for any subset $S \subset V$, such that $S \cap V_\alpha \neq \varnothing$ and $V_\alpha \setminus S \neq \varnothing$, consider an arbitrary nodes $v \in V_\alpha \cap S$ and $w \in V_\alpha \setminus S$. By construction, $\min\{\bar{y}_v, \bar{y}_w\} \geq (\alpha+1)/(\alpha+2)$. Therefore,

$$\hat{x}(\delta(S)) = \left(\frac{\alpha+2}{\alpha}\right) \bar{x}(\delta(S)) \geq 2\left(\frac{\alpha+2}{\alpha}\right)(\bar{y}_v + \bar{y}_w - 1)$$

$$\geq 2\left(\frac{\alpha+2}{\alpha}\right) \times \left[2\left(\frac{\alpha+1}{\alpha+2}\right) - 1\right] \geq 2.$$

Finally, relying on the following simple inequality:

$$1 - \hat{y}_w \leq (\alpha + 2)(1 - \bar{y}_w)$$

we obtain the desired upper bound

$$\frac{\mathrm{cost}(\hat{W})}{\mathrm{PCATSP_{LP}}^*} = \frac{c(x') + \sum\limits_{w \in V} \pi_w(1 - \hat{y}_w)}{\sum\limits_{a \in A} c_a \bar{x}_a + \sum\limits_{w \in V} \pi_v(1 - \bar{y}_w)} \leq \frac{\alpha \cdot \mathrm{ATSP_{LP}}^* + \sum\limits_{w \in V} \pi_w(1 - \hat{y}_w)}{\sum\limits_{a \in A} c_a \bar{x}_a + \sum\limits_{w \in V} \pi_v(1 - \bar{y}_w)}$$

$$\leq \frac{\alpha \cdot \sum\limits_{a \in A} c_a \hat{x}_a + \sum\limits_{w \in V} \pi_w(1 - \hat{y}_w)}{\sum\limits_{a \in A} c_a \bar{x}_a + \sum\limits_{w \in V} \pi_v(1 - \bar{y}_w)} \leq \alpha + 2.$$

Theorem is proved.

As a simple consequence of Theorem 2, we obtain

Remark 1. Employing the Svensson-Traub algorithm as Algorithm $\mathcal{A}_0$, we obtain $(24 + \varepsilon)$-approximation polynomial time algorithm for the PCATSP.

6   Conclusion

In this paper, the first approximation algorithm with constant ratio $(24+\varepsilon)$ was proposed for the asymmetric version of the Prize-Collecting Traveling Salesman Problem. Our algorithm owes its appearance to the breakthrough result by O. Svensson and V. Traub for the ATSP, which opened up the possibility to design high-accuracy approximation algorithms for asymmetric routing problems. We hope to present some of them including the algorithm for the general Balas formulation of the PCATSP in the near forthcoming paper.

References

1. Archer, A., Bateni, M., Hajiaghayi, M., Karloff, H.: Improved approximation algorithms for prize-collecting Steiner tree and tsp. SIAM J. Comput. **40**(2), 309–332 (2011). https://doi.org/10.1137/090771429

2. Balas, E.: The prize collecting traveling salesman problem. Networks **19**(6), 621–636 (1989). https://doi.org/10.1002/net.3230190602

3. Bateni, M., Chekuri, C., Ene, A., Hajiaghayi, M., Korula, N., Marx, D.: Prize-collecting steiner problems on planar graphs. In: Proceedings of the 2011 Annual ACM-SIAM Symposium on Discrete Algorithms (SODA), pp. 1028–1049 (2011). https://doi.org/10.1137/1.9781611973082.79

4. Bérubé, J.F., Gendreau, M., Potvin, J.Y.: A branch-and-cut algorithm for the undirected prize collecting traveling salesman problem. Networks **54**(1), 56–67 (2009). https://doi.org/10.1002/net.20307

5. Bienstock, D., Goemans, M.X., Simchi-Levi, D., Williamson, D.: A note on the prize collecting traveling salesman problem. Math. Program. **59**, 413–420 (1993). https://doi.org/10.1007/BF01581256

6. Chan, T.H.H., Jiang, H., Jiang, S.H.C.: A unified PTAS for prize collecting TSP and steiner tree problem in doubling metrics. In: Azar, Y., Bast, H., Herman, G. (eds.) 26th Annual European Symposium on Algorithms (ESA 2018). Leibniz International Proceedings in Informatics (LIPIcs), vol. 112, pp. 15:1–15:13. Schloss Dagstuhl-Leibniz-Zentrum fuer Informatik, Dagstuhl, Germany (2018). https://doi.org/10.4230/LIPIcs.ESA.2018.15

7. Christofides, N.: Worst-case analysis of a new heuristic for the traveling salesman problem. In: Symposium on New Directions and Recent Results in Algorithms and Complexity, p. 441 (1975)

8. Chung, S.H., Sah, B., Lee, J.: Optimization for drone and drone-truck combined operations: a review of the state of the art and future directions. Comput. Oper. Res. **123**, 105004 (2020). https://doi.org/10.1016/j.cor.2020.105004

9. Climaco, G., Simonetti, L., Rosetti, I.: A branch-and-cut and MIP-based heuristics for the prize-collecting travelling salesman problem. RAIRO-Oper. Res. **55**, S719–S726 (2021). https://doi.org/10.1051/ro/2020002

10. Dell'Amico, M., Maffioli, F., Värbrand, P.: On prize-collecting tours and the asymmetric travelling salesman problem. Int. Trans. Oper. Res. **2**(3), 297–308 (1995). https://doi.org/10.1016/0969-6016(95)00010-5

11. Dogan, O., Alkaya, A.F.: A novel method for prize collecting traveling salesman problem with time windows. In: Kahraman, C., Cebi, S., Cevik Onar, S., Oztaysi, B., Tolga, A.C., Sari, I.U. (eds.) INFUS 2021. LNNS, vol. 307, pp. 469–476. Springer, Cham (2022). https://doi.org/10.1007/978-3-030-85626-7_55

12. Fischetti, M., Toth, P.: Vehicle routing: methods and studies, chap. In: An Additive Approach for the Optimal Solution of the Prize Collecting Traveling Salesman Problem, pp. 319–343, Elsevier (1988)

13. Goemans, M.X.: Combining approximation algorithms for the prize-collecting tsp (2009). https://doi.org/10.48550/ARXIV.0910.0553, https://arxiv.org/abs/0910.0553

14. Goemans, M.X., Williamson, D.P.: A general approximation technique for constrained forest problems. SIAM J. Comput. **24**(2), 296–317 (1995). https://doi.org/10.1137/S0097539793242618

15. Gutin, G., Punnen, A.P.: The Traveling Salesman Problem and Its Variations. Springer, Boston (2007). https://doi.org/10.1007/b101971

16. Jackson, B.: Some remarks on arc-connectivity, vertex splitting, and orientation in graphs and digraphs. J. Graph Theory **12**(3), 429–436 (1988). https://doi.org/10.1002/jgt.3190120314

17. Lahyani, R., Khemakhem, M., Semet, F.: A unified matheuristic for solving multi-constrained traveling salesman problems with profits. EURO J. Comput. Optim. **5**(3), 393–422 (2016). https://doi.org/10.1007/s13675-016-0071-1

18. Lovász, L.: On some connectivity properties of Eulerian graphs. Acta Math. Acad. Sci. Hung. **28**(1), 129–138 (1976). https://doi.org/10.1007/BF01902503
19. Medeiros, Y.A.d., Goldbarg, M.C., Goldbarg, E.F.G.: Prize collecting traveling salesman problem with ridesharing. Revista de Informática Teórica e Aplicada 27(2), 13–29 (2020). https://doi.org/10.22456/2175-2745.94082
20. Nguyen, V.H., Nguyen, T.T.T.: Approximating the asymmetric profitable tour. Electr. Notes Discrete Math. **36**, 907–914 (2010). https://doi.org/10.1016/j.endm.2010.05.115
21. Papadimitriou, C.: Euclidean TSP is NP-complete. Theoret. Comput. Sci. **4**, 237–244 (1977)
22. Pedro, O., Saldanha, R., Camargo, R.: A tabu search approach for the prize collecting traveling salesman problem. Electr. Notes Discrete Math. **41**, 261–268 (2013). https://doi.org/10.1016/j.endm.2013.05.101
23. Rahbari, M., Jahed, A., Tehrani, N.S.: A hybrid simulated annealing algorithm for the prize collecting travelling salesman problem. In: 3rd International Conference on Industrial and Systems Engineering (ICISE 2017). Proceedings of 3rd International Conference on Industrial Engineering and Systems (ICISE 2017) (2017), https://hal.archives-ouvertes.fr/hal-01962060
24. Sahni, S., Gonzales, T.: P-complete approximation problems. J. ACM **23**, 555–565 (1976)
25. Svensson, O., Tarnawski, J., Végh, L.A.: A constant-factor approximation algorithm for the asymmetric traveling salesman problem. In: Proceedings of the 50th Annual ACM SIGACT Symposium on Theory of Computing, pp. 204–213. STOC 2018, Association for Computing Machinery, New York, NY, USA (2018). https://doi.org/10.1145/3188745.3188824
26. Traub, V., Vygen, J.: An Improved Approximation Algorithm for ATSP, In: STOC 2020: 52nd Annual ACM SIGACT Symposium on Theory of Computing, pp. 1–13. Association for Computing Machinery, New York, NY, USA (2020). https://doi.org/10.1145/3357713.3384233
27. Vansteenwegen, P., Gunawan, A.: Orienteering Problems: Models and Algorithms for Vehicle Routing Problems with Profits. Springer Cham (2019). https://doi.org/10.1007/978-3-030-29746-6

Optimal Control

Control of the Motion of Heating Sources of a Rod with Non-linear Feedback

K. R. Aida-zade[(✉)] [ID] and V.A. Hashimov[ID]

Institute of Control Systems of ANAS, B.Vahabzade 9, AZ1141 Baku, Azerbaijan
kamil_aydazade@rambler.ru
https://www.isi.az

Abstract. The problem of synthesis the control of the motion of lumped sources is studied on the example of the problem of control with feedback by moving heat sources when the rod is heated. The speeds of point sources are assigned depending on the state of the process at the measurement points. Formulas for the components of the functional gradient are obtained, which make it possible to use first-order optimization methods for the numerical solution of the problem.

Keywords: Rod heating · Feedback control · Moving sources · Temperature measurement points · Feedback parameters

1 Introduction

We study the problem of synthesis of control of the rod heating process by point-wise sources moving along the rod. The current values of the speeds of movement of the sources are determined depending on the temperature at the points of measurement. The paper proposes to use the non-linear dependence of the control actions by the motion of the sources on the measured temperature values. Constant coefficients involved in these dependencies are the desired feedback parameters.

Note that the problems of synthesis of control of objects described by both ordinary and partial differential equations are the most difficult both in the theory of optimal control and in the practice of their application [1–7].

For the problems of synthesis of control of objects with lumped parameters, there are certain, fairly general approaches to their solution, in particular, for linear systems [2–4,8]. There are no such approaches for objects with distributed parameters [1,3,5,6]. Firstly, this is due, to a wide variety of both mathematical models of such objects and possible variants of the corresponding formulations of control problems [1,3]. Secondly, the implementation of currently known methods for controlling objects with feedback in real time requires the use of expensive telemechanics, measuring and computing equipment [3,5,6].

In this paper, the determination of the feedback parameters is reduced to the problem of parametric optimal control. For the numerical solution of the problem, first-order optimization methods were used. Formulas for the gradient of the objective functional with respect to the feedback parameters are obtained.

N. Olenev et al. (Eds.): OPTIMA 2022, LNCS 13781, pp. 93–107, 2022.
https://doi.org/10.1007/978-3-031-22543-7_7

The proposed approach to the synthesis of the control of moving sources can be easily extended to other processes described by other types of differential equations and initial-boundary conditions.

2 Formulation of the Problem

Consider the following process of heating a rod by moving heat sources [5]:

$$u_t(x,t) = a^2 u_{xx}(x,t) - \lambda_0[u(x,t) - \theta] + \sum_{i=1}^{N_c} q_i \delta\left(x - z_i(t)\right), \tag{1}$$

$$x \in (0,l), \quad t \in (0,T],$$

$$u_x(0,t) = \lambda_1(u(0,t) - \theta), \quad t \in (0,T], \tag{2}$$
$$u_x(l,t) = -\lambda_2(u(l,t) - \theta), \quad t \in (0,T],$$

Here $u(x,t)$ is the temperature of the rod at the point x at the moment of time t; $\delta(\cdot)$ is a Dirac function; l is rod length; T is heating process duration; N_c is number of point sources; a, λ_0, λ_1, λ_2 are given coefficients; q_i are given heat source power, $z_i(t)$ are functions that determine the i-th rule of motion of source on the rod; moreover

$$0 \le z_i(t) \le l, \quad t \in [0,T], \quad i = 1,2,\ldots,N_c. \tag{3}$$

θ is constant temperature of the external environment with the known set of possible values Θ and the density function $\rho_\Theta(\theta)$ such that

$$\rho_\Theta(\theta) \ge 0, \quad \theta \in \Theta, \quad \int_\Theta \rho_\Theta(\theta)d\theta = 1.$$

With respect to the initial temperature of the rod the set of its possible values are known. This set is determined by a parametrically defined function:

$$u(x,0) = u_0(x;p), \quad x \in [0,l], \quad p \in P \subset \mathbf{R}^s. \tag{4}$$

where P is a given set of parameter values with a density function $\rho_P(p) \ge 0$

$$\rho_P(p) \ge 0, \quad p \in P, \quad \int_P \rho_P(p)dp = 1.$$

The motions of the sources are determined by the equations

$$\dot{z}(t) = a_i z_i(t) + \vartheta_i(t), \quad t \in (0,T], \tag{5}$$

$$z_i(0) = z_i^0, \quad i = 1,2,\ldots,N_c. \tag{6}$$

Here a_i are the given parameters of the source motion; z_i^0 are given initial positions of sources; $\vartheta_i(t)$ is a piece-wise continuous control, satisfying the conditions:

$$V = \left\{ \vartheta_i(t) : \quad \underline{\vartheta_i} \leq \vartheta_i(t) \leq \overline{\vartheta_i}, \quad i = 1, 2, \dots, N_c \right\}, \quad t \in [0, T]. \tag{7}$$

The problem of control is to determine the vector-function $\vartheta = \vartheta(t) = (\vartheta_1(t), \vartheta_2(t), \dots, \vartheta_{N_c}(t))$, minimizing the given functional:

$$J(\vartheta) = \int\limits_P \int\limits_\Theta I(\vartheta; p, \theta) \rho_\Theta(\theta) \rho_P(p) d\theta dp, \tag{8}$$

$$I(\vartheta; p, \theta) = \int\limits_0^l \mu(x)[u(x, T) - U(x)]^2 dx + \varepsilon \left\| \vartheta(t) - \hat{\vartheta} \right\|^2_{L_2^{N_c}[0,T]}. \tag{9}$$

Here $U(x)$, $\mu(x) \geq 0$, $x \in [0, l]$ are a given piece-wise continuous functions; $u(x, t) = u(x, t; \vartheta, p, \theta)$ is a solution to the initial-boundary value problem (1), (2), (3) under given control $\vartheta(t) \in V$, parameters $p \in P$ of initial condition $u_0(x; p)$ and temperature of the external environment $\theta \in \Theta$. $\varepsilon > 0$, $\hat{\vartheta}$ are given parameters of regularization of the functional of the problem.

The objective functional in the problem under consideration estimates the control vector function $\vartheta(t)$ on the behavior of the heating process on average over all possible values of the parameters of the initial conditions $p \in P$ and temperature of the external environment $\theta \in \Theta$.

Let at the given N_o points $\xi_j \in [0, l]$, $j = 1, 2, \dots, N_o$ continuously temperature is measured:

$$\check{u}_j(t) = u(\xi_j, t), \quad t \in [0, T], \quad \xi_j \in [0, l], \quad j = 1, 2, \dots, N_o.$$

It is required to control the movement of sources, taking into account the results of current temperature measurements and the distance between sources and point measurements. For this we use the following non-linear relationship:

$$\vartheta_i(t) = \sum_{j=1}^{N_o} \left(\alpha_i^j \left(z_i(t) - \xi_j \right)^2 + \beta_i^j \right) [u(\xi_j, t) - \gamma_i^j], \quad t \in [0, T], \tag{10}$$

$$i = 1, 2, \dots, N_c.$$

The expression in the first parenthesis is determined by the distance between the measurement point and the current source location. The second bracket is determined by the difference between the measured temperature values and the optimized nominal values at the temperature measurement points on the rod.

Let us transform the feedback (10) to the form:

$$\vartheta_i(t) = \sum_{j=1}^{N_o} \left[y_1^{ij} z_i^2(t) u(\xi_j, t) + y_2^{ij} z_i(t) u(\xi_j, t) + y_3^{ij} u(\xi_j, t) \right] \tag{11}$$

$$+y_4^i z_i^2(t) + y_5^i z_i(t) + y_6^i, \quad t \in [0,T], \quad i = 1,2,\ldots,N_c.$$

where the notation for the feedback parameters is used: $y_1^{ij} = \alpha_i^j$, $y_2^{ij} = -2\alpha_i^j \xi_j$, $y_3^{ij} = \alpha_i^j \xi_j^2 + \beta_i^j$, $y_4^i = -\sum_{j=1}^{N_o} \alpha_i^j \gamma_i^j$, $y_5^i = 2\sum_{j=1}^{N_o} \alpha_i^j \gamma_i^j \xi_j$, $y_6^i = -\sum_{j=1}^{N_o} (\alpha_i^j \xi_j^2 + \beta_i^j)\gamma_i^j$.

Introducing the notation for $N = 3N_c(N_o + 1)$ dimensional vector $y = (y_1^{ij}, y_2^{ij}, y_3^{ij}, y_4^i, y_5^i, y_6^i)$. The objective functional in this case can be written as:

$$J(y) = \int_P \int_\Theta I(y; p, \theta)\rho_\Theta(\theta)\rho_P(p)d\theta dp, \tag{12}$$

$$I(y; p, \theta) = \int_0^l \mu(x)[u(x,T) - U(x)]^2 dx + \varepsilon \|y - \hat{y}\|_{R^N}^2. \tag{13}$$

Substituting dependencies (11) into equations (5), we obtain

$$\dot{z}_i(t) = a^i z_i(t) + \sum_{j=1}^{N_o} \left[y_1^{ij} z_i^2(t)u\left(\xi_j, t\right) + y_2^{ij} z_i(t)u\left(\xi_j, t\right) + y_3^{ij} u\left(\xi_j, t\right) \right] \tag{14}$$

$$+y_4^i z_i^2(t) + y_5^i z_i(t) + y_6^i, \quad t \in [0,T], \quad i = 1,2,\ldots,N_c.$$

From technological considerations, the range of possible temperature values at the points of the rod during its heating can be considered known:

$$\underline{u} \le u(x,t) \le \overline{u}, \quad x \in [0,l], \quad t \in [0,T]. \tag{15}$$

Then, taking into account the linearity of dependences (11), from constraints (7) we obtain linear constraints on the feedback parameters:

$$\underline{\vartheta}_i \le \sum_{j=1}^{N_o} \left[y_1^{ij} z_i^2(t) + y_2^{ij} z_i(t) + y_3^{ij} \right] \tilde{u}^k + y_4^i z_i^2(t) + y_5^i z_i(t) + y_6^i \le \overline{\vartheta}_i, \tag{16}$$

$$i = 1,2,\ldots,N_c, \quad k = 1,2,\ldots,\tilde{N}.$$

Here $\tilde{u}^k$ is the j-th column of the matrix $\tilde{U}$ of size $N_o \times \tilde{N}$, $\tilde{N} = 2^{N_o}$. The columns of the matrix $\tilde{U}$ are N_o dimensional vectors consisting of various combinations of values $\underline{u}, \overline{u}$.

Thus, the source control problem for moving sources (1)–(9) with feedback (11) is reduced to the parametric optimal control problem (12), (13), (1), (2), (3) [7,9].

We note the following specific features of the investigated parametric optimal control problem.

First, the original problem of control of moving sources (1)–(9) is generally not convex. This is due to the non-linear third term parameters in equation (1) $\delta(x - z_i(t; \vartheta_i(t)))$. The resulting problem of parametric optimal control is also not convex in the feedback parameters y. This can be seen from the differential equations (14) and dependences (11).

Secondly, the problem is specific because of the objective functional (12), (13), which estimates the behavior of a beam of phase trajectories with initial conditions from a parametrically given set.

In general, the obtained problem can also be classified as a class of finite-dimensional optimization problems with respect to the vector $y \in \mathrm{R}^N$. In this problem, to calculate the objective functional at any point, it is required to solve the initial-boundary value problem with respect to the differential equation with partial derivatives (1) and the system of differential equations with ordinary derivatives (5)

3 Determination of Feedback Parameters

To minimize the functional (12), (13), taking into account the linearity of constraints (16), we use the gradient projection method [9]:

$$y^{n+1} = \mathcal{P}_{(16)}\left[y^n - \alpha_n \mathbf{grad} J\left(y^n\right)\right], \tag{17}$$

$$\alpha_n = \arg \min_{\alpha \geq 0} J\left(\mathcal{P}_{(16)}\left[y^n - \alpha \mathbf{grad} J\left(y^n\right)\right]\right), \quad n = 0, 1, \dots$$

Here α_n is the one-dimensional minimization step, y^0 is an arbitrary starting point of the search from R^N; $\mathcal{P}_{(16)}[\cdot]$ is the operator of projecting an arbitrary point $y \in \mathrm{R}^N$ onto the admissible domain defined by constraints (16). Taking into account the linearity of constraints (16), the operator $\mathcal{P}_{(16)}[\cdot]$ is easy to construct constructively [9]. It is known that iterative procedure (17) allows one to find only the local minimum of the objective functional closest to the point y^0. Therefore, for procedure (17), it is proposed to use the multistart method from different starting points. From the obtained local minimum points, the best functional is selected.

In the implementation of procedure (17), analytical formulas for the components of the gradient of the objective functional play an important role. Therefore, below we will prove the differentiability of the functional with respect to the optimized parameters and obtain formulas for its gradient, which make it possible to formulate the necessary optimal conditions for the synthesized feedback parameters y.

Theorem 1. *Under conditions on the functions and parameters involved in the problem (1), (2), (4), (6), (12)–(14), the functional (12), (13) is differentiable and components of its gradient with respect to the feedback parameters are determined by the formulas:*

$$\frac{\partial J(y)}{\partial y_1^{ij}} = \int\limits_P \int\limits_\Theta \left\{ -\int\limits_0^T \varphi_i(t) z_i^2(t) u\left(\xi_j, t\right) dt + 2\varepsilon(y_1^{ij} - \hat{y}_1^{ij}) \right\} \rho_\Theta(\theta) \rho_P(p) d\theta dp,$$

$$\frac{\partial J(y)}{\partial y_2^{ij}} = \int\limits_P \int\limits_\Theta \left\{ -\int\limits_0^T \varphi_i(t) z_i(t) u\left(\xi_j, t\right) dt + 2\varepsilon(y_2^{ij} - \hat{y}_2^{ij}) \right\} \rho_\Theta(\theta) \rho_P(p) d\theta dp,$$

$$\frac{\partial J(y)}{\partial y_3^{ij}} = \int\limits_P \int\limits_\Theta \left\{ -\int\limits_0^T \varphi_i(t) u\left(\xi_j, t\right) dt + 2\varepsilon(y_3^{ij} - \hat{y}_3^{ij}) \right\} \rho_\Theta(\theta)\rho_P(p)d\theta dp,$$

$$\frac{\partial J(y)}{\partial y_4^i} = \int\limits_P \int\limits_\Theta \left\{ -\int\limits_0^T \varphi_i(t) z_i^2(t) dt + 2\varepsilon(y_4^i - \hat{y}_4^i) \right\} \rho_\Theta(\theta)\rho_P(p)d\theta dp, \qquad (18)$$

$$\frac{\partial J(y)}{\partial y_5^i} = \int\limits_P \int\limits_\Theta \left\{ -\int\limits_0^T \varphi_i(t) z_i(t) dt + 2\varepsilon(y_5^i - \hat{y}_5^i) \right\} \rho_\Theta(\theta)\rho_P(p)d\theta dp,$$

$$\frac{\partial J(y)}{\partial y_6^i} = \int\limits_P \int\limits_\Theta \left\{ -\int\limits_0^T \varphi_i(t) dt + 2\varepsilon(y_6^i - \hat{y}_6^i) \right\} \rho_\Theta(\theta)\rho_P(p)d\theta dp,$$

$i = 1, 2, \ldots, N_c,\; j = 1, 2, \ldots, N_o.$ *Functions* $\psi(x,t)$ *and* $\vartheta_i(t)$, $i = 1, 2, \ldots, N_c$ *for every given parameters* $\theta \in \Theta$ *and* $p \in P$ *are solutions of the following conjugate initial-boundary value problems:*

$$\psi_t(x,t) = -a^2\psi_{xx}(x,t) + \lambda_0\psi(x,t)$$

$$-\sum_{j=1}^{N_o} \delta\left(x - \xi_j\right) \sum_{i=1}^{N_c} \varphi_i(t) \left[y_1^{ij} z_i^2(t) + y_2^{ij} z_i(t) + y_3^{ij} \right], \quad x \in (0, l), \quad t \in [0, T),$$

$$\psi(x, T) = -2\mu(x)\left(u(x,T) - U(x)\right), \quad x \in [0, l], \qquad (19)$$

$$\psi_x(0, t) = \lambda_1\psi(0, t), \quad t \in [0, T),$$

$$\psi_x(l, t) = -\lambda_2\psi(l, t), \quad t \in [0, T),$$

$$\dot{\varphi}_i(t) = -a_i\varphi_i(t) - \psi_x\left(z_i(t), t\right)q_i \qquad (20)$$

$$-\varphi_i(t) \left\{ \sum_{j=1}^{N_o} \left[2y_1^{ij} z_i(t) u\left(\xi_j, t\right) + y_2^{ij} u\left(\xi_j, t\right) \right] - 2y_4^i z_i(t) - y_5^i \right\}, \quad t \in [0, T),$$

$$\varphi_i(T) = 0, \quad i = 1, 2, \ldots, N_c.$$

Proof. To prove the differentiability of the functional $J(y)$ with respect to y, we use the increment method.

From the obvious dependence of all parameters of the initial conditions $p \in P$ and the temperature of the external environment $\theta \in \Theta$, the validity of the formula follows:

$$\mathbf{grad}J(y) = \mathbf{grad} \int\limits_P \int\limits_\Theta I(y; p, \theta)\rho_\Theta(\theta)\rho_P(p)d\theta dp \qquad (21)$$

$$= \int\limits_P \int\limits_\Theta \mathbf{grad}I(y; p, \theta)\rho_\Theta(\theta)\rho_P(p)d\theta dp.$$

Therefore, it is sufficient to obtain formulas for $\mathbf{grad}I(y; p, \theta)$ for arbitrarily given admissible values of $p \in P$ and $\theta \in \Theta$. On the right side of equation (1), we introduce the notation

$$W(t; y) = \sum_{i=1}^{N_c} q_i \delta \left(x - z_i(t) \right), \quad t \in (0, T].$$

Let $u(x, t) = u(x, t; y, p, \theta)$, $z(t) = z(t; y, p, \theta)$ are solutions, respectively, initial boundary-value problem (1), (2), (4) and Cauchy problem (14), (6) for given values of the parameters p and θ. Let the feedback parameters y get an increment Δy: $\tilde{y} = y + \Delta y$. It is clear that the corresponding solutions of problems (1), (2), (4) and (14), (6) also get increments, which we denote as follows:

$$\tilde{u}\left(x, t; \tilde{y}, p, \theta \right) = u(x, t; y, p, \theta) + \Delta u(x, t; y, p, \theta),$$

$$\tilde{z}_i\left(t; \tilde{y}, p, \theta \right) = z_i(t; y, p, \theta) + \Delta z_i(t; y, p, \theta), \quad i = 1, 2, \ldots, N_c.$$

The increments $\Delta u(x, t; y, p, \theta)$ and $\Delta z_i(t; y, p, \theta)$, $i = 1, 2, \ldots, N_c$ are solutions of the following initial-boundary-value problems with an accuracy of small above the first order of with respect to $\|\Delta u(x, t)\|$, $\|\Delta z(t)\|$, $\|\Delta y\|$:

$$\Delta u_t(x, t) = a^2 \Delta u_{xx}(x, t) - \lambda_0 \Delta u(x, t) + \Delta W(t; y), \quad x \in (0, l), \quad t \in (0, T],$$
$$(22)$$

$$\Delta u(x, 0) = 0, \quad x \in [0, l], \tag{23}$$

$$\Delta u_x(0, t) \doteq \lambda_1 \Delta u(0, t), \quad t \in (0, T], \tag{24}$$

$$\Delta u_x(l, t) = -\lambda_2 \Delta u(l, t), \quad t \in (0, T].$$

$$\Delta \dot{z}(t) = a_i \Delta z_i(t) + \Delta \vartheta_i(t), \quad t \in (0, T], \tag{25}$$

$$\Delta z_i(0) = 0, \quad i = 1, 2, \ldots, N_c. \tag{26}$$

The functional $\Delta I(y; p, \theta)$ will get an increment

$$\Delta I(y) = I(y + \Delta y; p, \theta) - I(y; p, \theta) \tag{27}$$

$$= 2 \int_0^l \mu(x) \left(u(x, T) - U(x) \right) \Delta u(x, T) dx + 2\varepsilon \langle y - \widehat{y}, \Delta y \rangle + \mathcal{R}_1,$$

$$\mathcal{R}_1 = o \left(\|\Delta u(x, t)\|, \|\Delta y\| \right).$$

Move the right-hand sides of differential Eqs. (22) and (25) to the left, multiply both sides of the obtained qualities by so far arbitrary functions $\psi(x, t)$ and $\varphi_i(t)$, respectively. We integrate over $t \in (0, T)$ and $x \in (0, l)$. The resulting left-hand sides equal to zero are added to (27). Will have:

$$\Delta I(y) = 2 \int_0^l \mu(x) \left(u(x, T) - U(x) \right) \Delta u(x, T) dx + 2\varepsilon \langle y - \hat{y}, \Delta y \rangle$$

$$+ \int\limits_0^T \int\limits_0^l \psi(x,t) \left(\Delta u_t(x,t) - a^2 \Delta u_{xx}(x,t) + \lambda_0 \Delta u(x,t) - \Delta W(t;y) \right) dxdt$$

$$+ \sum_{i=1}^{N_c} \int\limits_0^T \varphi_i(t) \left(\Delta z_i(t) - a_i \Delta z_i(t) - \Delta \vartheta_i(t) \right) dt + \mathcal{R}_1.$$

Integrating by parts, grouping and taking into account conditions (23), (24), (26), we obtain the following expression for the increment of the functional:

$$\Delta I(y) = 2 \int\limits_0^l \mu(x) \left(u(x,T) - U(x) \right) \Delta u(x,T) dx + \int\limits_0^l \psi(x,T) \Delta u(x,T) dx \quad (28)$$

$$+ \int\limits_0^T \int\limits_0^l \left(-\psi_t(x,t) - a^2 \psi_{xx}(x,t) + \lambda_0 \psi(x,t) \right) \Delta u(x,t) dxdt$$

$$+ a^2 \int\limits_0^T \left(\psi_x(l,t) + \lambda_2 \psi(l,t) \right) \Delta u(l,t) dt - a^2 \int\limits_0^T \left(\psi_x(0,t) - \lambda_1 \psi(0,t) \right) \Delta u(0,t) dt$$

$$- \sum_{i=1}^{N_c} \sum_{j=1}^{N_o} \int\limits_0^T \int\limits_0^l \varphi_i(t) \left(y_1^{ij} z_i^2(t) + y_2^{ij} z_i(t) + y_3^{ij} \right) \delta\left(x - \xi_j \right) \Delta u(x,t) dxdt$$

$$+ \sum_{i=1}^{N_c} \varphi_i(T) \Delta z_i(T) + \sum_{i=1}^{N_c} \int\limits_0^T \left\{ -\dot{\varphi}_i(t) - a_i \varphi_i(t) - \psi_x\left(z_i(t),t \right) q_i \right\} \Delta z_i(t) dt$$

$$- \sum_{i=1}^{N_c} \sum_{j=1}^{N_o} \int\limits_0^T \left\{ \varphi_i(t) \left(2 y_1^{ij} z_i(t) u\left(\xi_j,t \right) + y_2^{ij} u\left(\xi_j,t \right) + 2 y_4^i z_i(t) + y_5^i \right) \Delta z_i(t) \right\} dt$$

$$+ \sum_{i=1}^{N_c} \sum_{j=1}^{N_o} \Delta y_1^{ij} \left\{ -\int\limits_0^T \varphi_i(t) z_i^2(t) u\left(\xi_j,t \right) dt + 2\varepsilon(y_1^{ij} - \hat{y}_1^{ij}) \right\}$$

$$+ \sum_{i=1}^{N_c} \sum_{j=1}^{N_o} \Delta y_2^{ij} \left\{ -\int\limits_0^T \varphi_i(t) z_i(t) u\left(\xi_j,t \right) dt + 2\varepsilon(y_2^{ij} - \hat{y}_2^{ij}) \right\}$$

$$+ \sum_{i=1}^{N_c} \sum_{j=1}^{N_o} \Delta y_3^{ij} \left\{ -\int\limits_0^T \varphi_i(t) u\left(\xi_j,t \right) dt + 2\varepsilon(y_3^{ij} - \hat{y}_3^{ij}) \right\}$$

$$+ \sum_{i=1}^{N_c} \sum_{j=1}^{N_o} \Delta y_4^i \left\{ -\int\limits_0^T \varphi_i(t) z_i^2(t) dt + 2\varepsilon(y_4^i - \hat{y}_4^i) \right\}$$

$$+ \sum_{i=1}^{N_c} \sum_{j=1}^{N_o} \Delta y_5^i \left\{ - \int_0^T \varphi_i(t) z_i(t) dt + 2\varepsilon(y_5^i - \hat{y}_5^i) \right\}$$

$$+ \sum_{i=1}^{N_c} \sum_{j=1}^{N_o} \Delta y_6^i \left\{ - \int_0^T \varphi_i(t) dt + 2\varepsilon(y_6^i - \hat{y}_6^i) \right\} + \mathcal{R}_2,$$

$$\mathcal{R}_2 = o\left(\|\Delta u(x,t)\|, \|\Delta z(t)\|, \|\Delta y\| \right).$$

Using the well-known results on the solution of the boundary value problem (1), (2) and the Cauchy problem (5), (6), one can obtain estimates $\|\Delta u(x;t)\| < k_1\|\Delta y\|$, $\|\Delta u(x,t)\| < k_2\|\Delta y\|$. Then from (28) it follows that the functional of the problem is differentiable.

Using the arbitrariness of the choice of the functions $\psi(x,t)$ and $\varphi_i(t)$, we require them to satisfy conditions (19) and (20).

Then the components of the gradient of the functional $I(y; p, \theta)$, determined by the linear parts of the functional increment with the corresponding feedback parameters [9], are defined by the following formulas:

$$\frac{\partial I(y; p, \theta)}{\partial y_1^{ij}} = - \int_0^T \varphi_i(t) z_i^2(t) u(\xi_j, t) dt + 2\varepsilon(y_1^{ij} - \hat{y}_1^{ij}),$$

$$\frac{\partial I(y; p, \theta)}{\partial y_2^{ij}} = - \int_0^T \varphi_i(t) z_i(t) u(\xi_j, t) dt + 2\varepsilon(y_2^{ij} - \hat{y}_2^{ij}),$$

$$\frac{\partial I(y; p, \theta)}{\partial y_3^{ij}} = - \int_0^T \varphi_i(t) u(\xi_j, t) dt + 2\varepsilon(y_3^{ij} - \hat{y}_3^{ij}), \tag{29}$$

$$\frac{\partial I(y; p, \theta)}{\partial y_4^i} = - \int_0^T \varphi_i(t) z_i^2(t) dt + 2\varepsilon(y_4^i - \hat{y}_4^i),$$

$$\frac{\partial I(y; p, \theta)}{\partial y_5^i} = - \int_0^T \varphi_i(t) z_i(t) dt + 2\varepsilon(y_5^i - \hat{y}_5^i),$$

$$\frac{\partial I(y; p, \theta)}{\partial y_6^i} = \int_0^T \varphi_i(t) dt + 2\varepsilon(y_6^i - \hat{y}_6^i).$$

Taking into account formula (21) from (29), we obtain the required formulas (18) given in Theorem 1.

4   Numerical Experiments

The purpose of the numerical experiments carried out to solve test problems of the form (1)–(9) with feedback (10) was as follows. We studied 1) the quality of the objective functional, namely the property of multi-extremality; 2) dependence of the solution of the control problem on the number of sources and points for measuring the state; 3) the influence of the state measurement error on the control of the heating process as a whole.

Numerical experiments were carried out on the example of test problem, in which the parameters and functions involved in (1)–(9) were as follows:

$$a^2 = 1, \quad \lambda_0 = 0.001, \quad \lambda_1 = \lambda_2 = 0.0001, \quad l = 1, \quad T = 1, \quad q_1 = q_2 = 5,$$

$$\underline{\vartheta_1} = \underline{\vartheta_2} = -2, \quad \overline{\vartheta_1} = \overline{\vartheta_2} = 2, \quad z_1(t) \in [0.05; 0.95], \quad z_2(t) \in [0.05; 0.95],$$

$$\xi_1 = 0.2, \quad \xi_2 = 0.4, \quad \xi_3 = 0.6, \quad \xi_4 = 0.8,$$

$$N_\varphi = N_\theta = 3, \quad \Phi = \{0.4; 0.6; 0.8\}, \quad \Theta = \{4.8; 5; 5.2\},$$

$$U(x) = 10, \quad x \in [0,1], \quad P(\varphi = \varphi_i) = P(\theta = \theta_j) = 1/3, \quad i, j = 1, \dots, 3.$$

When minimizing the objective functional, the method of external penalty functions was used to take into account restrictions (3) and (15), and the gradient projection method (17) was used to take into account restrictions (16) [9]. To find the step α at each iteration of (17), the golden section method was applied [9].

To solve the direct and conjugate boundary value problems of parabolic type, an implicit scheme of the grid method was used: with a step in the spatial variable $h_x = 0.01$, and for the time variable $h_t = 0.001$ [10, 11].

To approximate the function and Dirac, taking into account that the sources are moving, the scheme proposed in [12] was used.

First, consider the case when the number of sources $N_c = 2$ and the number of measurement points $N_o = 4$. Tables 1 and 2 show the results of some intermediate iterations of the objective function gradient projection method obtained for two different starting points for procedure (17).

As can be seen from the tables, different initial values of feedback parameters y were used for these experiments.

Note that as a result of solving the problem from different initial points for minimization, the obtained values of the feedback parameters differ quite strongly. At the same time, the obtained minimum values of the objective functional are almost the same.

Figure 1a and b) shows the initial and optimal source trajectories for different initial approximations y^0.

For comparison, the case was also considered when the power of the sources is equal to $q_1 = q_2 = 10$, and the duration of the control time is equal to $T = 0.5$.

Tables 3, 4 show the results of intermediate iterations from different initial points used for the procedure (17). Figure 2. shows the source motion trajectories for the initial and obtained values of the optimized parameters. As can be

Table 1. The results of solution of the problem for $N_c = 2$, $N_o = 4$, $q_1 = q_2 = 5$, $T = 1$ from the first starting point.

N	y								J(y)
0	0.022478	0.028745	0.021258	0.021289	-0.021478	-0.022815	-0.020058	-0.021125	0.840330
	0.029785	0.021248	0.029874	0.022887	-0.023569	-0.023587	-0.021489	-0.022589	
	0.022548	0.023598	0.022158	0.021258	-0.022987	-0.021258	-0.023745	-0.021587	
	0.038752	-0.035587	0.029862	-0.028698	0.012458	-0.011458			
1	0.008475	0.014742	0.007255	0.007286	-0.014517	-0.015854	-0.013097	-0.014164	0.051825
	0.012680	0.004143	0.012769	0.005782	-0.009590	-0.009608	-0.007510	-0.008610	
	0.001200	0.002250	0.000810	-0.000090	0.012990	0.014719	0.012232	0.014390	
	0.037337	-0.033331	0.028122	-0.025254	0.010244	-0.004929			
2	0.008526	0.014793	0.007306	0.007337	-0.013013	-0.014350	-0.011593	-0.012660	0.01064
	0.013174	0.004637	0.013263	0.006276	-0.007682	-0.007700	-0.005602	-0.006702	
	0.006352	0.007402	0.005962	0.005062	0.015422	0.017151	0.014664	0.016822	
	0.037345	-0.032987	0.028214	-0.024838	0.011368	-0.004423			
3	0.008525	0.014792	0.007305	0.007336	-0.013660	-0.014997	-0.012240	-0.013307	0.007812
	0.013170	0.004633	0.013259	0.006272	-0.008469	-0.008487	-0.006389	-0.007489	
	0.006305	0.007355	0.005915	0.005015	0.014462	0.016191	0.013704	0.015862	
	0.037345	-0.033121	0.028213	-0.024996	0.011348	-0.004610			

Table 2. The results of solution of the problem for $N_c = 2$, $N_o = 4$, $q_1 = q_2 = 5$, $T = 1$ from the second starting point.

N	y								J(y)
0	0.016048	0.027048	0.022048	0.014048	-0.017512	-0.013512	-0.008512	-0.009512	0.6505
	0.018414	0.014414	0.035414	0.033414	-0.021436	-0.024436	-0.013436	-0.011436	
	0.005922	0.012922	0.012922	0.015922	-0.032346	-0.031346	-0.024346	-0.023346	
	0.012007	-0.022869	0.020073	-0.020854	0.015837	-0.015836			
1	0.006045	0.007045	0.002045	0.004045	-0.008377	-0.004377	-0.001377	-0.001377	0.0198
	0.008392	0.004392	0.005392	0.003392	-0.002443	-0.005443	-0.004443	-0.002443	
	0.004727	0.008727	0.012727	0.005727	-0.003521	-0.002521	-0.005521	-0.004521	
	0.010007	-0.003073	0.020068	-0.001082	0.005787	-0.006093			
2	0.006043	0.007043	0.002043	0.004043	-0.008323	-0.004323	-0.001323	-0.001323	0.0091
	0.008392	0.004392	0.005392	0.003392	-0.002329	-0.005329	-0.004329	-0.002329	
	0.004790	0.008790	0.012790	0.005790	-0.003322	-0.002322	-0.005322	-0.004322	
	0.010006	-0.003050	0.020068	-0.001049	0.005796	-0.006047			
3	0.006047	0.007047	0.002047	0.004047	-0.008105	-0.004105	-0.001105	-0.001105	0.007843
	0.008421	0.004421	0.005421	0.003421	-0.002050	-0.005050	-0.004050	-0.002050	
	0.005035	0.009035	0.013035	0.006035	-0.002964	-0.001964	-0.004964	-0.003964	
	0.010007	-0.002991	0.020073	-0.000980	0.005848	-0.005964			

Table 3. The results of solution of the problem for $N_c = 2$, $N_o = 4$, $q_1 = q_2 = 10$, $T = 0.5$ from the first starting point.

N	y								J(y)
0	0.022478	0.028745	0.021258	0.021289	-0.021478	-0.022815	-0.020058	-0.021125	0.9635
	0.029785	0.021248	0.029874	0.022887	-0.023569	-0.023587	-0.021489	-0.022589	
	0.022548	0.023598	0.022158	0.021258	-0.022987	-0.021258	-0.023745	-0.021587	
	0.038752	-0.035587	0.029862	-0.028698	0.012458	-0.011458			
1	0.022791	0.029058	0.021571	0.021602	-0.014029	-0.015366	-0.012609	-0.013676	0.4076
	0.031509	0.022972	0.031598	0.024611	-0.012643	-0.012661	-0.010563	-0.011663	
	0.033913	0.034963	0.033523	0.032623	-0.006314	-0.004585	-0.007072	-0.004914	
	0.038798	-0.033273	0.030168	-0.025750	0.015151	-0.007561			
2	0.020869	0.027136	0.019649	0.019680	-0.011442	-0.012779	-0.010022	-0.011089	0.2073
	0.024704	0.016167	0.024793	0.017806	-0.009080	-0.009098	-0.007000	-0.008100	
	0.002795	0.003845	0.002405	0.001505	-0.001359	0.000370	-0.002117	0.000041	
	0.038535	-0.032770	0.029052	-0.025112	0.008029	-0.006739			
3	0.020878	0.027145	0.019658	0.019689	-0.010239	-0.011576	-0.008819	-0.009886	0.0292
	0.025533	0.016996	0.025622	0.018635	-0.007502	-0.007520	-0.005422	-0.006522	
	0.017029	0.018079	0.016639	0.015739	0.000710	0.002439	-0.000048	0.002110	
	0.038539	-0.032274	0.029220	-0.024525	0.011073	-0.006040			

Table 4. The results of solution of the problem for $N_c = 2$, $N_o = 4$, $q_1 = q_2 = 10$, $T = 0.5$ from the second starting point.

N	y								J(y)
0	0.006000	0.007000	0.002000	0.004000	-0.008000	-0.004000	-0.001000	-0.001000	0.2696
	0.008000	0.004000	0.005000	0.003000	-0.002000	-0.005000	-0.004000	-0.002000	
	0.001000	0.005000	0.009000	0.002000	-0.003000	-0.002000	-0.005000	-0.004000	
	0.010000	-0.003000	0.020000	-0.001000	0.005000	-0.006000			
0	0.006063	0.007063	0.002063	0.004063	-0.013764	-0.009764	-0.006764	-0.006764	0.0740
	0.008808	0.004808	0.005808	0.003808	-0.008432	-0.011432	-0.010432	-0.008432	
	0.011684	0.015684	0.019684	0.012684	-0.010185	-0.009185	-0.012185	-0.011185	
	0.010011	-0.004428	0.020151	-0.002560	0.007240	-0.007707			
2	0.006157	0.007157	0.002157	0.004157	-0.012296	-0.008296	-0.005296	-0.005296	0.0530
	0.009469	0.005469	0.006469	0.004469	-0.006620	-0.009620	-0.008620	-0.006620	
	0.017039	0.021039	0.025039	0.018039	-0.007932	-0.006932	-0.009932	-0.008932	
	0.010024	-0.003959	0.020261	-0.002020	0.008381	-0.007080			
3	-0.034613	-0.033613	-0.038613	-0.036613	-0.043340	-0.039340	-0.036340	-0.036340	0.0294
	-0.042348	-0.046348	-0.045348	-0.047348	-0.034819	-0.037819	-0.036819	-0.034819	
	0.090872	0.094872	0.098872	0.091872	-0.037695	-0.036695	-0.039695	-0.038695	
	0.004630	-0.015537	0.017089	-0.014084	0.059046	-0.019976			

Table 5. Results of solution the problem for $N_c = 4$, $N_o = 4$, $q_i = 2.5$, $i = 1, 2, \ldots, N_c$, $T = 1$.

N	y										J(y)
0	0.0059	0.0069	0.0019	0.0039	−0.0042	−0.0002	−0.0028	−0.0028	−0.0090	−0.0080	0.5798
	−0.0050	−0.0150	−0.0199	−0.0129	−0.0099	−0.0099	0.0086	0.0046	0.0056	0.0036	
	0.0050	0.0020	0.0025	0.0045	−0.0093	−0.0133	−0.0073	−0.0173	−0.0118	−0.0148	
	−0.0138	−0.0148	0.0094	0.0034	0.0174	0.0024	0.0041	0.0051	0.0085	0.0056	
	−0.0432	−0.0392	0.0458	−0.0242	−0.0015	−0.0005	−0.0035	−0.0025	0.0099	0.0084	
	−0.0041	−0.0079	0.0201	0.0084	0.0082	−0.0099	0.0514	−0.0574	0.0145	−0.0093	
1	0.0063	0.0073	0.0023	0.0043	−0.0036	0.0003	−0.0022	−0.0022	−0.0088	−0.0078	0.0327
	−0.0049	−0.0149	−0.0195	−0.0125	−0.0095	−0.0095	0.0099	0.0059	0.0069	0.0049	
	0.0056	0.0026	0.0032	0.0051	−0.0088	−0.0128	−0.0068	−0.0168	−0.0106	−0.0136	
	−0.0126	−0.0136	0.0145	0.0085	0.0225	0.0075	0.0047	0.0057	0.0091	0.0062	
	−0.0416	−0.0376	0.0474	−0.0226	0.0014	0.0024	−0.0006	0.0004	0.0100	0.0091	
	−0.0041	−0.0078	0.0203	0.0085	0.0083	−0.0096	0.0526	−0.0572	0.0149	−0.0090	
2	0.0059	0.0069	0.0019	0.0039	−0.0058	−0.0018	0.0012	0.0012	0.0087	0.0047	0.0005
	0.0057	0.0037	−0.0032	−0.0062	−0.0052	−0.0032	0.0077	0.0038	0.0047	0.0028	
	0.0006	−0.0024	−0.0014	0.0006	0.0024	0.0063	0.0104	0.0033	−0.0053	−0.0043	
	−0.0073	−0.0063	−0.0012	0.0028	0.0068	−0.0002	0.0001	0.0011	−0.0019	−0.0009	
	0.0027	0.0027	0.0027	0.0027	−0.0044	−0.0044	−0.0044	−0.0044	0.0100	−0.0025	
	0.0201	−0.0013	0.0199	−0.0005	0.0053	−0.0065	0.0045	−0.0054	0.0006	−0.0009	

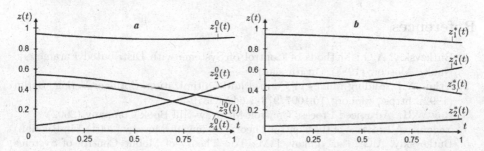

Fig. 1. Trajectories of sources for different initial minimization parameters (- - - initial trajectory, − optimal trajectory)

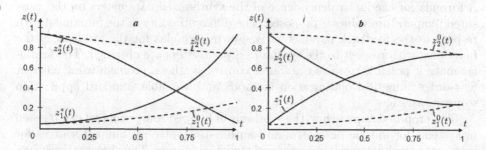

Fig. 2. Trajectories of motion of sources for different initial (a) and obtained (b) feedback parameters.

seen from the tables, despite the fact that the obtained minimum values of the objective functional are the same, the optimized parameters are different.

Table 5 shows the results of solving the problem with the number of sources $N_c = 4$ and the number of measurement points $N_o = 4$. Duration of control process is $T = 1$, and the powers of the sources are $q_i = 2.5$, $i = 1, \ldots, 4$.

An increase in the number of sources made it possible to significantly reduce the value of the objective functional. Figure 2 shows the initial and optimal trajectories of the sources.

Numerous calculations of the heating process were carried out at optimal values of the feedback parameters, at which the state measurements had an error of 2%, 3%, 5%. The calculations showed that both the deviations of the trajectories of the sources and the course of the heating process itself do not differ significantly from the corresponding indicators with accurate measurements. Due to their small difference, it is not possible to present comparative results in a visual form.

5   Conclusion

In the article the problem of synthesis of motion control of lumped sources of heating of the rod is investigated. The rules of motion of pointwise sources are assigned depending on the state of the process at the measurement points. A formula for the linear dependence of the synthesized parameters on the measured temperature values is proposed. The differentiability of the functional with respect to the feedback parameters is shown, formulas for the gradient of the functional with respect to the synthesized parameters are obtained. The formulas make it possible to solve the source control synthesis problem using efficient first-order numerical optimization methods and available standard application software packages.

The proposed approach to the synthesis of lumped source control can be used in systems for automatic control and automatic control of lumped sources for many other technological processes and technical objects. The objects themselves can be described by other partial differential equations and types of boundary conditions.

References

1. Butkovskiy, A.G.: Methods of Control of Systems with Distributed Parameters. Nauka, Moscow (1984). (In Russian)
2. Utkin, V.I.: Sliding Modes in Control and Optimization. Springer-Verlag, Berlin (1992). https://doi.org/10.1007/978-3-642-84379-2
3. Ray W.H.: Advanced Process Control. McGraw-Hill Book Company (2002)
4. Yegorov, A.I.: Bases of the Control Theory. Fizmatlit, Moscow (2004). (In Russian)
5. Butkovskiy, A.G., Pustylnikov, L.M.: The Theory of Mobile Control of Systems with Distributed Parameters. Nauka, Moscow (1980). (In Russian)
6. Sirazetdinov, T.K.: Optimization of Systems with Distributed Parameters. Nauka, Moscow (1977). (In Russian)

7. Sergienko, I.V., Deineka, V.S.: Optimal Control of Distributed Systems with Conjugation Conditions. Kluwer Acad. Publisher, New York (2005)
8. Polyak, B.T., Khlebnikov, M.V., Rapoport, L.B.: Mathematical Theory of Automatic Control. LENAND, Moscow (2019). (In Russian)
9. Vasilyev, F.P.: Optimization methods, 824p. Faktorial Press, Moscow (2002). (In Russian)
10. Abdullaev, V.M., Aida-Zade, K.R.: Numerical method of solution to loaded non-local boundary value problems for ordinary differential equations, Comp. Math. Math. Phys. **54**(7), 1096–1109 (2014)
11. Abdullayev, V.M., Aida-zade, K.R.: Finite-difference methods for solving loaded parabolic equations. Comp. Math. Math. Phys. **56**(1), 93–105 (2016)
12. Aida-zade, K.R., Hashimov, V.A., Bagirov, A.H.: On a problem of synthesis of control of power of the moving sources on heating of a rod. Proc. Inst. Math. Mech. ANAS **47**(1), 183–196 (2021)

Terminal Control of Multi-agent System

Anatoly Antipin[1]([✉])[iD] and Elena Khoroshilova[2]([✉])[iD]

[1] FRC "Computer Science and Control" RAS, Vavilov 40, 119333 Moscow, Russia
asantip@yandex.ru
[2] Lomonosov MSU, CMC Faculty, Leninskiye Gory, 119991 Moscow, Russia
khorelena@gmail.com

Abstract. A linear controlled dynamics is considered on a fixed time interval. Dynamics transforms control into a phase trajectory. This trajectory at discrete points of the time interval is loaded with finite-dimensional linear programming problems. These problems define intermediate, initial and boundary value solutions, which correspond to the ends of time subsegments. It is required by the choice of control to form a phase trajectory so that, starting from the initial conditions, the trajectory passes through all solutions of intermediate problems and reaches the terminal conditions at the right end of the time interval. In general, constructions that combine dynamics with mathematical programming problems will be called terminal control problems. The approach to solving these problems is based on the Lagrangian formalism and duality theory. The paper proposes an iterative saddle point computing process for solving the problem of terminal control, which belongs to the class of multi-agent systems. The study was carried out within the framework of evidence-based methodology, i.e. the convergence of computational process with respect to all components of solution is proved.

Keywords: Terminal control · Lagrangian formalism · Duality ·
Intermediate problems · Linear programming · Saddle point methods ·
Evidence-based optimization · Convergence

1 Statement of the Control Problem

We consider a linear differential control system that, on a fixed time interval $[t_0, t_S]$, transforms the control into a phase trajectory defined on the same interval. The situation is complicated by the fact that a large time segment is divided into S subsegments that touch each other at their ends

$$\Gamma = \{t_0, t_1, ..., t_{s-1}, t_s, t_{s+1}, ..., t_S\}.$$

The sampling points, on the one hand, are the ends of the time subsegments on which the differential control system is projected (or narrowed). On the other hand, at these points the phase trajectory is loaded with linear programming problems.

© The Author(s), under exclusive license to Springer Nature Switzerland AG 2022
N. Olenev et al. (Eds.): OPTIMA 2022, LNCS 13781, pp. 108–120, 2022.
https://doi.org/10.1007/978-3-031-22543-7_8

Thus, on each of subsegments $[t_{s-1}, t_s]$, $s = \overline{1,S}$, its own section of the phase trajectory is defined, which we denote as $x_s(t)$. At common points of adjacent segments $[t_{s-1}, t_s]$ and $[t_s, t_{s+1}]$, the values of the trajectories coincide by construction: $x_s(t_s) = x_{s+1}(t_s)$, i.e., on each of the subsegments, the right end of the trajectory coincides with the starting point of the trajectory on the next subsegment.

As a result of such discretization of the time interval, we arrive at the following statement of multi-agent problem:

$$
\begin{cases}
\dfrac{d}{dt}x(t) = D(t)x(t) + B(t)u(t), \quad t \in [t_0, t_S], \ x(t) \in AC^n[t_0, t_S], \ u(t) \in U, \\
\qquad x(t_0) = x^0, \ x(t_1) = x_1^*, \ x(t_2) = x_2^*, \ \dots, \ x(t_S) = x_S^*, \\
\quad x_1^* \in \mathrm{Argmin}\{\langle \varphi_1, x_1 \rangle \mid G_1 x_1 \le g_1, \ x_1 \in \mathbb{R}^n\}, \ x_1^* \in X_1, \\
\quad x_2^* \in \mathrm{Argmin}\{\langle \varphi_2, x_2 \rangle \mid G_2 x_2 \le g_2, \ x_2 \in \mathbb{R}^n\}, \ x_2^* \in X_2, \\
\qquad \cdots\cdots\cdots\cdots\cdots\cdots\cdots\cdots\cdots\cdots\cdots\cdots\cdots\cdots\cdots\cdots \\
\quad x_S^* \in \mathrm{Argmin}\{\langle \varphi_S, x_S \rangle \mid G_S x_S \le g_S, \ x_S \in \mathbb{R}^n\}, \ x_S^* \in X_S.
\end{cases}
$$

$$(1)$$

Here $x_s(t_s)$ are the values of functions $x_s(t)$ on subsegment $[t_{s-1}, t_s]$ calculated at the right end of the s-th subsegment, x_s^* is a solution to the s-th intermediate linear programming problem; φ_s are the normals of the objective functions; X_s are intermediate reachable sets; $G_s = G(t_s)$, $g_s = g(t_s)$, $s = \overline{1,S}$, there are given matrices and vectors of phase constraints; U is a convex closed set of admissible controls. Combining the functions $x_s(t)$, $s = \overline{1,S}$, we obtain a trajectory $x(t)$, $t \in [t_0, t_S]$, on the entire segment. In other words, we "cut" the original problem (1) into S independent problems of the same kind. This partition has a fractal character.

We note one more circumstance. In each of the intermediate finite-dimensional optimization problems, the minimum of the linear function is sought at the intersection of the finite-dimensional polytope $\{G_s x_s \le g_s, \ x_s \in \mathbb{R}^n\}$ and the reachable set $X_s \subset \mathbb{R}^n$. However, the reachable set is not included in the problem constraints that describe the feasible set. This is due to the fact that in the considered approach the right end of the phase trajectory automatically belongs to the admissible set.

For greater clarity, we present system (1) in expanded form. Γ-sampling generates time intervals $[t_{s-1}, t_s]$ on which functions $x_s(t)$ are defined for all $s = \overline{1,S}$ (functions are indexed by right ends of their segments). Each of these functions is a narrowing of phase trajectory $x(t)$ to segment $[t_{s-1}, t_s]$. In this model, on each s-th time segment $[t_{s-1}, t_s]$, the s-th controlled trajectory $x_s(t)$ and the s-th intermediate problem are defined:

$$
\begin{cases}
\dfrac{d}{dt}x_1(t) = D(t)x_1(t) + B(t)u_1(t), \ t \in [t_0, t_1], \\
\qquad x_1(t_0) = x^0, \ x_1(t_1) = x_1^*, \ u_1(t) \in U, \\
x_1^* \in \mathrm{Argmin}\{\langle \varphi_1, x_1 \rangle \mid G_1 x_1 \le g_1, \ x_1 \in \mathbb{R}^n\}, \ x_1^* \in X_1, \ x_1(t_1) = x_1,
\end{cases}
$$

$$\left\{ \begin{array}{c} \dfrac{d}{dt}x_s(t) = D(t)x_s(t) + B(t)u_s(t), \ t \in [t_{s-1}, t_s], \\[2mm] x_s(t_{s-1}) = x^*_{s-1}, \ x_s(t_s) = x^*_s, \ u_s(t) \in \mathrm{U}, \\[2mm] x^*_s \in \mathrm{Argmin}\{\langle \varphi_s, x_s \rangle \mid G_s x_s \le g_s, \ x_s \in \mathbb{R}^n\}, \ x^*_s \in X_s, \ x_s(t_s) = x_s, \end{array} \right. \quad (2)$$

$$\left\{ \begin{array}{c} \dfrac{d}{dt}x_S(t) = D(t)x_S(t) + B(t)u_S(t), \ t \in [t_{S-1}, t_S], \\[2mm] x_S(t_{S-1}) = x^*_{S-1}, \ x_S(t_S) = x^*_S, \ u_S(t) \in \mathrm{U}, \\[2mm] x^*_S \in \mathrm{Argmin}\{\langle \varphi_S, x_S \rangle \mid G_S x_S \le g_S, \ x_S \in \mathbb{R}^n\}, \ x^*_S \in X_S, \ x_S(t_s) = x_s. \end{array} \right.$$

So, within the framework of the proposed approach, the source problem (1) is split into a finite set of independent intermediate terminal control problems. Each of these problems can be solved independently, starting with the first. Then, by drawing a phase trajectory through solutions of intermediate problems, one can find a solution to the terminal control problem on entire segment $[t_0, t_S]$. To solve any of the subproblems of system (1), methods have been developed in [2,3].

2   Problem Statement in Vector-Matrix Form

System (2) is very convenient to be represented in vector-matrix format:

dynamics

$$\begin{pmatrix} \frac{dx_1}{dt} \\ \frac{dx_2}{dt} \\ \vdots \\ \frac{dx_S}{dt} \end{pmatrix} = \begin{pmatrix} D_1 & 0 & \cdots & 0 \\ 0 & D_2 & \cdots & 0 \\ \vdots & \vdots & \ddots & \vdots \\ 0 & 0 & \cdots & D_S \end{pmatrix} \begin{pmatrix} x_1 \\ x_2 \\ \vdots \\ x_S \end{pmatrix} + \begin{pmatrix} B_1 & 0 & \cdots & 0 \\ 0 & B_2 & \cdots & 0 \\ \vdots & \vdots & \ddots & \vdots \\ 0 & 0 & \cdots & B_S \end{pmatrix} \begin{pmatrix} u_1 \\ u_2 \\ \vdots \\ u_S \end{pmatrix},$$

where $x(t_0) = x_0$, $x_s(t_s) = x^*_s$, $x_S(t_S) = x^*_S$, $u_s(t) \in \mathrm{U}$, and

intermediate problems

$$\begin{pmatrix} x^*_1 \\ x^*_2 \\ \vdots \\ x^*_S \end{pmatrix} \in \mathrm{Argmin} \left\{ (\varphi_1 \ \varphi_2 \cdots \varphi_S) \begin{pmatrix} x_1 \\ x_2 \\ \vdots \\ x_S \end{pmatrix} \Bigg| \begin{pmatrix} G_1 & 0 & \cdots & 0 \\ 0 & G_2 & \cdots & 0 \\ \vdots & \vdots & \ddots & \vdots \\ 0 & 0 & \cdots & G_S \end{pmatrix} \begin{pmatrix} x_1 \\ x_2 \\ \vdots \\ x_S \end{pmatrix} \le \begin{pmatrix} g_1 \\ g_2 \\ \vdots \\ g_S \end{pmatrix} \right\} \quad (3)$$

Recall that each function $x(t)$ generates vector $x = (x(t_1), ..., x(t_S))$ whose number of components is equal to the number of discretization points on $[t_0, t_S]$. Each component of x, in turn, is a vector of size n. Thus, we have the space $\mathbb{R}^{n \times S}$. In this space, a diagonal matrix $G(t_s)$ is defined with submatrices $G_s(t_s)$, $s = \overline{1, S}$, from (3). We have described a matrix functional constraint of the type inequality with the right side, which is given by the vector $g = (g_1, g_2, ..., g_S)$. The linear objective function that completes the formulation of the finite-dimensional linear programming problem in (3) is the scalar product of vectors φ and x.

In macro format, we can represent problem (3) as

$$\begin{cases} \dfrac{d}{dt}x(t) = D(t)x(t) + B(t)u(t), \ t_0 \le t \le t_S, \ x(t_0) = x^0, \ x(t_S) = x_S^*, \\ \qquad x^* \in \text{Argmin}\{\langle \varphi, x \rangle \mid Gx \le g, \ x \in \mathbb{R}^n\}, \ u(t) \in \text{U}, \end{cases} \quad (4)$$

where $D(t)$, $B(t)$, $G(t)$ are continuous matrices of size $n \times n$, $n \times r$, $m \times n$ respectively; $g(t)$ is a given continuous vector function; $G_S = G(t_S)$, $g_S = g(t_S)$, $x_S = x(t_S)$ are the matrix and vector values at the right end of the time segment; φ is a given vector (normal to a linear objective functional), $x^*(t_0) = x^0$ is a given initial condition. The controls $u(t)$ for each $t \in [t_0, t_S]$ belong to the set U, which is a convex closed set of $\mathbb{R}^r$.

Note that macrosystem (4), which is obtained as a result of scalarization of intermediate problems (3), or (4), almost completely coincides in format with a terminal control problem with a boundary value problem at the right end proposed and studied in [2,3]. Therefore, the method for solving problem (4) and the proof of its convergence in general will repeat the logic of reasoning from [2,3].

As a solution to differential system (4) we mean any pair $(x(t), u(t)) \in L_2^n[t_0, t_S] \times \text{U}$ satisfying identically the following condition

$$x(t) = x(t_0) + \int_{t_0}^{t} (D(\tau)x(\tau) + B(\tau)u(\tau))d\tau, \quad t_0 \le t \le t_S. \quad (5)$$

The identity defines a generalized solution of dynamics (4). It is shown in [14] that any control $u(t) \in \text{U}$ in the linear differential system corresponds to a single trajectory $x(t)$, and this pair satisfies identity (5). In applications, the control $u(t)$ is often a piecewise continuous function. Moreover, the presence of discontinuity points on control $u(t)$ has no effect on the values of trajectory $x(t)$. Moreover, the trajectory will remain unchanged even if we change the values of function $u(t)$ on a set of measure zero. The trajectory $x(t)$ in situation (5) is absolutely continuous function. The class of absolutely continuous functions is a linear variety dense in $L_2^n[t_0, t_S]$. In what follows, this class will be denoted as $AC^n[t_0, t_S] \subset L_2^n[t_0, t_S]$. For any pair of functions $(x(t), u(t)) \in AC^n[t_0, t_S] \times \text{U}$, the Newton-Leibniz formulas and, accordingly, the integration-by-parts formulas are satisfied. It is assumed that solutions $(x^*(t), u^*(t)) \in AC^n[t_0, t_S] \times \text{U}$ exist [13,14].

3   Classical Lagrangian

The considered problem is a linear programming problem formulated in Hilbert space. In the theory of linear programming in a finite-dimensional space, it is known that along with the primal problem, there is always a dual problem in the dual space. By drawing appropriate analogies, one can try to explicitly obtain the dual problem for system (4). To this end, we scalarize system (4) and introduce a linear convolution known as the Lagrange function

$$\mathcal{L}(p, \psi(t); x, x(t), u(t)) = \langle \varphi, x \rangle + \langle p, Gx - g \rangle$$

$$+ \int_{t_0}^{ts} \langle \psi(t), D(t)x(t) + B(t)u(t) - \frac{d}{dt}x(t) \rangle dt, \tag{6}$$

for all p, $\psi(t)$, x, $x(t), u(t)$ of consistent dimensions, where $\psi(t)$ belongs to the linear variety of absolutely continuous functions from the dual space.

The saddle point $(p^*, \psi^*(t); x^*, x^*(t), u^*(t))$ of the Lagrange function is formed by primal variables $(x^*, x^*(t), u^*(t))$ and dual variables $(p^*, \psi^*(t))$. These variables are solutions to problems (1)–(4) and, by definition, satisfy the following system of inequalities

$$\langle \varphi, x^* \rangle + \langle p, Gx^* - g \rangle + \int_{t_0}^{ts} \langle \psi(t), D(t)x^*(t) + B(t)u^*(t) - \frac{d}{dt}x^*(t) \rangle dt$$

$$\leq \langle \varphi, x^* \rangle + \langle p^*, Gx^* - g \rangle + \int_{t_0}^{ts} \langle \psi^*(t), D(t)x^*(t) + B(t)u^*(t) - \frac{d}{dt}x^*(t) \rangle dt$$

$$\leq \langle \varphi, x \rangle + \langle p^*, Gx - g_1 \rangle + \int_{t_0}^{ts} \langle \psi^*(t), D(t)x(t) + B(t)u(t) - \frac{d}{dt}x(t) \rangle dt \tag{7}$$

for all p, $\psi(t)$, x, $x(t), u(t)$.

So, if the source problem (1)–(4) has primal and dual solutions, then this pair forms a saddle point of the Lagrange function. Here, as in the finite-dimensional case, the dual solution is the normal to the linear functional, which is the support functional at the minimum point that satisfies the linear constraints (2).

Further, we will show that the converse statement is true: the saddle point of the Lagrange function (6) is formed by primal and dual solutions of source problems (1)–(4).

The left-hand inequality of system (7) is the problem of maximizing a linear function with respect to variables $(p, \psi(t))$ over the entire space of variation of these variables:

$$\langle p - p^*, Gx^* - g \rangle + \int_{t_0}^{ts} \langle \psi(t) - \psi^*(t), D(t)x^*(t) + B(t)u^*(t) - \frac{d}{dt}x^*(t) \rangle dt \leq 0. \tag{8}$$

It follows from inequality (8) that

$$\langle p - p^*, Gx^* - g \rangle \leq 0,$$

$$D(t)x^*(t) + B(t)u^*(t) - \frac{d}{dt}x^*(t) = 0, \quad x^*(t_0) = x^0 \tag{9}$$

for all p. Setting first $p = 0$ and then $p = 2p^*$, we get

$$\langle p^*, Gx^* - g \rangle = 0, \quad Gx^* - g \leq 0,$$

$$D(t)x^*(t) + B(t)u^*(t) - \frac{d}{dt}x^*(t) = 0, \quad x^*(t_0) = x^0. \tag{10}$$

The right-hand inequality of system (7) is the problem of minimizing the Lagrange function with respect to variables $x, x(t), u(t)$ for fixed values $p = p^*$,

$\psi(t) = \psi^*(t)$. We show that vectors $(p^*, \psi^*(t); x^*, x^*(t), u^*(t))$ form a solution to (1)–(4). Taking into account (10), from the right-hand inequality of system (7) we have

$$\langle \varphi, x^* \rangle \leq \langle \varphi, x \rangle + \langle p^*, Gx - g \rangle + \int_{t_0}^{t_S} \langle \psi^*(t), D(t)x(t) + B(t)u(t) - \frac{d}{dt}x(t) \rangle dt \quad (11)$$

for all $x, x(t), u(t)$.

Consider inequality (11) under additional scalar constraints

$$\langle p^*, Gx - g \rangle \leq 0, \quad \int_{t_0}^{t_S} \langle \psi^*(t), D(t)x(t) + B(t)u(t) - \frac{d}{dt}x(t) \rangle dt = 0.$$

Then we get the optimization problem

$$\langle \varphi, x^* \rangle \leq \langle \varphi, x \rangle$$

under constraints

$$\langle p^*, Gx - g \rangle \leq 0, \quad \int_{t_0}^{t_S} \langle \psi^*(t), D(t)x(t) + B(t)u(t) - \frac{d}{dt}x(t) \rangle dt = 0 \quad (12)$$

for all $x, x(t), u(t)$.

It follows from (10) that solution $(x^*(t), u^*(t))$ belongs to a narrower set than (12). Therefore, the indicated point remains a minimum on the subset of solutions to system (10), i.e.,

$$\langle \varphi, x^* \rangle \leq \langle \varphi, x \rangle, \quad Gx \leq g, \quad (13)$$

$$\frac{d}{dt}x(t) = D(t)x(t) + B(t)u(t) \quad (14)$$

for all $x, x(t), u(t)$. Thus, if the Lagrange function (6) has a saddle point, then its vector of primal components is a solution to the source problem of convex programming in an infinite-dimensional space.

4   Dual Lagrangian and Dual Problem

We will show that the Lagrange function plays the role of a "bridge", along which one can pass from the source problem in the primal space to the dual problem in the dual space.

Using formulas for passing to adjoint linear operators

$$\langle \psi, Dx \rangle = \langle D^T\psi, x \rangle, \quad \langle \psi, Bu \rangle = \langle B^T\psi, u \rangle \quad (15)$$

and formulas for integration by parts on segment $[t_0, t_S]$

$$\langle \psi(t_S), x(t_S) \rangle - \langle \psi(t_0), x(t_0) \rangle = \int_{t_0}^{t_S} \langle \frac{d}{dt}\psi(t), x(t) \rangle dt + \int_{t_0}^{t_S} \langle \psi(t), \frac{d}{dt}x(t) \rangle dt, \quad (16)$$

we write the Lagrange function conjugate with respect to (6) and the saddle point system (7) in the conjugate form:

$$\mathcal{L}^T(p, \psi(t); x, x(t), u(t)) = \langle \varphi + G^T p - \psi, x \rangle - \langle g, p \rangle + \langle \psi^0, x^0 \rangle$$

$$+ \int_{t_0}^{t_S} \langle D^T(t)\psi(t) + \frac{d}{dt}\psi(t), x(t) \rangle dt + \int_{t_0}^{t_S} \langle B^T(t)\psi(t), u(t) \rangle dt \qquad (17)$$

for all $p, \psi(t), x, x(t), u(t)$, $x^0 = x(t_0)$, $\psi^0 = \psi(t_0)$, $\psi_S = \psi(t_S)$.

Both Lagrangians (primal and dual) have the same saddle point $(p^*, \psi^*(t); x^*, x^*(t), u^*(t))$ that satisfies the saddle point conjugate (dual) system

$$\langle \varphi + G^T p - \psi, x^* \rangle + \langle -g, p \rangle + \langle \psi^0, x^0 \rangle$$

$$+ \int_{t_0}^{t_S} \langle D^T(t)\psi(t) + \frac{d}{dt}\psi(t), x^*(t) \rangle dt + \int_{t_0}^{t_S} \langle B^T(t)\psi(t), u^*(t) \rangle dt$$

$$\leq \langle \varphi + G^T p^* - \psi^*, x^* \rangle + \langle -g, p^* \rangle + \langle \psi^0, x^0 \rangle$$

$$+ \int_{t_0}^{t_S} \langle D^T(t)\psi^*(t) + \frac{d}{dt}\psi^*(t), x^*(t) \rangle dt + \int_{t_0}^{t_S} \langle B^T(t)\psi^*(t), u^*(t) \rangle dt$$

$$\leq \langle \varphi + G^T p^* - \psi^*, x \rangle + \langle -g, p^* \rangle + \langle \psi^0, x^0 \rangle$$

$$+ \int_{t_0}^{t_S} \langle D^T(t)\psi^*(t) + \frac{d}{dt}\psi^*(t), x(t) \rangle dt + \int_{t_0}^{t_S} \langle B^T(t)\psi^*(t), u(t) \rangle dt \qquad (18)$$

for all $p, \psi(t), x, x(t), u(t)$.

From the right-hand inequality (18) we have

$$\langle \varphi + G^T p^* - \psi^*, x^* - x \rangle + \int_{t_0}^{t_S} \langle D^T(t)\psi^*(t) + \frac{d}{dt}\psi^*(t), x^*(t) - x(t) \rangle dt$$

$$+ \int_{t_0}^{t_S} \langle B^T(t)\psi^*(t), u^*(t) - u(t) \rangle dt \leq 0$$

for all $x, x(t), u(t)$. For $u(t) = u^*(t)$, from the resulting inequality we have

$$\langle \varphi + G^T p^* - \psi^*, x^* - x \rangle + \int_{t_0}^{t_S} \langle D^T(t)\psi^*(t) + \frac{d}{dt}\psi^*(t), x^*(t) - x(t) \rangle dt \leq 0 \quad (19)$$

for all $x, x(t)$. For $x(t) = x^*(t)$ from here we get

$$\int_{t_0}^{t_S} \langle B^T(t)\psi^*(t), u^*(t) - u(t) \rangle dt \leq 0 \qquad (20)$$

for all $u(t)$.

Considering that (19) is the problem of maximizing a linear function on the entire space with respect to variables $x, x(t)$, the pair (19), (20) can be rewritten in the form

$$D^T(t)\psi^*(t) + \frac{d}{dt}\psi^*(t) = 0, \quad \varphi + G^T p^* - \psi^* = 0,$$

$$\int_{t_0}^{t_S} \langle B^T(t)\psi^*(t), u^*(t) - u(t)\rangle dt \leq 0. \tag{21}$$

From the left-hand side of (18), taking into account equations (21), we have

$$\langle \varphi + G^T p - \psi, x^*\rangle + \langle -g, p\rangle + \int_{t_0}^{t_S} \langle D^T(t)\psi(t) + \frac{d}{dt}\psi(t), x^*(t)\rangle dt$$

$$+ \int_{t_0}^{t_S} \langle B^T(t)\psi(t), u^*(t)\rangle dt \leq \langle -g, p^*\rangle + \int_{t_0}^{t_S} \langle B^T(t)\psi^*(t), u^*(t)\rangle dt.$$

Consider this inequality under a pair of scalar constraints

$$\langle \varphi + G^T p - \psi, x^*\rangle = 0,$$

$$\int_{t_0}^{t_S} \langle D^T(t)\psi(t) + \frac{d}{dt}\psi(t), x^*(t)\rangle dt = 0. \tag{22}$$

Then we get the problem of maximizing the scalar function under scalar constraints (22)

$$\langle -g, p\rangle + \int_{t_0}^{t_S} \langle \psi(t), B(t)u^*(t)\rangle dt \leq \langle -g, p^*\rangle + \int_{t_0}^{t_S} \langle \psi^*(t), B(t)u^*(t)\rangle dt$$

whence we arrive at the dual problem with respect to dual variables under vector constraints:

$$(p^*, \psi^*(t)) \in \text{Argmax}\{\langle -g, p\rangle + \int_{t_0}^{t_S} \langle \psi(t), B(t)u^*(t)\rangle dt \;\Big| \tag{23}$$

$$D^T(t)\psi(t) + \frac{d}{dt}\psi(t) = 0, \quad \psi = \varphi + G^T p\}. \tag{24}$$

Thus, we can see that (22), (23) and (20) form a dual problem with respect to (1)–(4).

5   Mutually Dual Problems

We write together a pair of mutually dual problems:
primal problem

$$x^* \in \text{Argmin}\{\langle \varphi, x\rangle \mid Gx \leq g,$$

$$\frac{d}{dt}x(t) = D(t)x(t) + B(t)u(t), \quad x(t_0) = x^0\}, \tag{25}$$

dual problem

$$(p^*, \psi^*(t)) \in \text{Argmax}\{\langle -g, p \rangle + \int_{t_0}^{t_S} \langle \psi(t), B(t)u^*(t) \rangle dt,$$

$$D^T(t)\psi(t) + \frac{d}{dt}\psi(t) = 0, \quad \psi = \varphi + G^T p\}, \tag{26}$$

$$\int_{t_0}^{t_S} \langle B^T(t)\psi^*(t), u^*(t) - u(t) \rangle dt \leq 0. \tag{27}$$

Each of problems (25)–(27) individually or in combination can become the basis for developing a whole family of methods for calculating the saddle point of primal or dual Lagrange functions [4–12,15]. In this case, it is possible to construct saddle point methods that will converge monotonically in the norm to saddle points of the Lagrange functions.

In this paper, we consider an iterative process for solving the boundary value differential system, which, on the one hand, will be obtained from the saddle point inequalities, and, on the other hand, will be close to the boundary value differential system, as if the latter were obtained from the condition of the maximum principle Pontryagin.

6   Sufficient Conditions for Extremality

We consider together the left-hand inequality (7) for the classical Lagrangian and the right-hand inequality (18) for the dual Lagrangian. Within these systems, subsystems (9) and (26) were obtained. Combining them together, we write out the general system

$$\frac{d}{dt}x^*(t) = D(t)x^*(t) + B(t)u^*(t), \quad x^*(t_0) = x^0,$$

$$\langle p - p^*, Gx^* - g \rangle \leq 0,$$

$$D^T(t)\psi^*(t) + \frac{d}{dt}\psi^*(t) = 0, \quad \psi^* = \varphi + G^T p^*,$$

$$\int_{t_0}^{t_S} \langle B^T(t)\psi^*(t), u^*(t) - u(t) \rangle dt \leq 0 \tag{28}$$

for all $p, u(t)$.

The terminal system (28) was obtained from the necessary and sufficient conditions for the saddle point of the Lagrange function. The variational inequalities of this system can be rewritten in the equivalent form of operator equations with projection operators onto the corresponding convex closed sets. Then we obtain a system of differential and operator equations of the form [13,14]

$$\frac{d}{dt}x^*(t) = D(t)x^*(t) + B(t)u^*(t), \quad x^*(t_0) = x^0, \tag{29}$$

$$p^* = \pi_+(p^* + \alpha(Gx^* - g)), \tag{30}$$

$$D^T(t)\psi^*(t) + \frac{d}{dt}\psi^*(t) = 0, \quad \psi^* = \varphi + G^T p^*, \tag{31}$$

$$u^*(t) = \pi_U(u^*(t) - \alpha B^T(t)\psi^*(t)), \tag{32}$$

where π_+, π_U are projection operators onto the positive space orthant for variable p and onto the set of controls, $\alpha > 0$. Here $(p^*, \psi^*(t), x^*, x^*(t), u^*(t))$ is the set of vectors that are solutions to (29)–(32).

7 Saddle Point Method Based on Sufficient Extremality Conditions

On the basis of (29)–(32), we construct an iterative process. If we take arbitrary values of dual variable $p = p^k$ and control $u(t) = u^k(t)$, which we treat as some approximations, then we can solve differential equation (29) and find the trajectory $x^k(t)$. Using p^k and $x^k(t)$, we implement the iteration (30). After that, using the transversality condition, we calculate terminal condition $\psi^k = \varphi + G^T p^k$ of conjugate system (31), solve it, and find the conjugate trajectory $\psi^k(t)$. Using the latter and control $u^k(t)$, we iterate over control (32) and find the next control with number $u^{k+1}(t)$.

Formally, the process has the form

$$\frac{d}{dt}x^k(t) = D(t)x^k(t) + B(t)u^k(t), \quad x^k(t) = x^0(t), \tag{33}$$

$$p^{k+1} = \pi_+(p^k + \alpha(Gx^k - g)), \tag{34}$$

$$D^T(t)\psi^k(t) + \frac{d}{dt}\psi^k(t) = 0, \quad \psi^k = \varphi + G^T p^k, \tag{35}$$

$$u^{k+1}(t) = \pi_U(u^k(t) - \alpha B^T(t)\psi^k(t); \quad k = 0, 1, 2, ... \tag{36}$$

Here, each next iteration is actually reduced to solving two systems of differential equations (33), (35).

Process (33)–(36) refers to simple iteration methods and is the simplest known computational process. However, in our case, we have a saddle point problem, about which it is known that methods of the simple iteration type do not converge to the saddle point (only their analogues in optimization converge, for example, gradient projection methods).

Therefore, we proposed an iterative saddle point (extragradient) method, the formulas of which are:

1) predictive half step

$$\frac{d}{dt}x^k(t) = D(t)x^k(t) + B(t)u^k(t), \quad x^k(t_0) = x^0, \tag{37}$$

$$\bar{p}^k = \pi_+(p^k + \alpha(Gx^k - g)), \tag{38}$$

$$\frac{d}{dt}\psi^k(t) + D^T(t)\psi^k(t) = 0, \quad \psi^k(t) = \varphi + G^T p^k, \tag{39}$$

$$\bar{u}^k(t) = \pi_U(u^k(t) - \alpha B^T(t)\psi^k(t)); \tag{40}$$

2) basic half step

$$\frac{d}{dt}\bar{x}^k(t) = D(t)\bar{x}^k(t) + B(t)\bar{u}^k(t), \quad \bar{x}^k(t_0) = x^k(t_0), \tag{41}$$

$$p^{k+1} = \pi_+(p^k + \alpha(G\bar{x}^k(t) - g)), \tag{42}$$

$$\frac{d}{dt}\bar{\psi}^k(t) + D^T(t)\bar{\psi}^k(t) = 0, \quad \bar{\psi}^k(t) = \varphi + G^T \bar{p}^k, \tag{43}$$

$$u^{k+1}(t) = \pi_U(u^k(t) - \alpha B^T(t)\bar{\psi}^k(t)), \quad k = 0, 1, 2, \ldots \tag{44}$$

Here, at each half step, two differential equations are solved and an iterative step over controls is performed. Note that in this process, iterations over primal variables $(x^k(t), u^k(t))$ for all k always belong to admissible sets and are solutions to differential equations (37), (39) and (41), (43). This process can be considered internal, or admissible, since each member of the iterative sequence always belongs to the admissible set.

It is proved that the process (37)–(44) of the extragradient method converges weakly in controls and strongly in primal and dual variables to one of solutions of the source problem.

The following convergence theorem for the proposed method was proved.

Theorem 1. *If the set of solutions $(p^*, \psi^*(t); x^*, x^*(t), u^*(t))$ for problem (29)–(32) is not empty, then sequence $(p^k, \psi^k(t); x^k, x^k(t), u^k(t))$ generated by method (37)-(44) with the step length $\alpha \leq \alpha_0$, where α_0 is some value, contains subsequence $(p^{k_i}, \psi^{k_i}(t); x^{k_i}, x^{k_i}(t), u^{k_i}(t))$, which converges to the solution of the problem, including: weak convergence in controls, strong convergence in trajectories, conjugate trajectories, and also in terminal variables. In particular, sequence $\{|p^k - p^*|^2 + \|u^k(t) - u^*(t)\|^2\}$ decreases monotonically.*

8 Conclusions

The article investigates the problem of terminal control of a multi-agent system. This problem belongs to the class of multi-agent linear systems and has a convex structure. The latter makes it possible, within the framework of the Lagrangian formalism and duality theory, to use the saddle point properties of the Lagrangian. In particular, this makes it possible to develop the saddle point theory of gradient methods in finite-dimensional intermediate spaces for solving multi-agent terminal control problems. The proposed approach combines the iterative recalculation of the saddle point flow of gradient methods in intermediate finite-dimensional spaces with the ability to "pull along" graphs of continuous phase trajectories. In the limit, we obtain that the limit points of iterative processes are solutions to finite-dimensional convex programming problems and lie on the graph of the phase trajectory, which, in turn, is a solution to the optimal control problem.

Another important point is that all calculations were carried out in the framework of evidence-based methodology. In other words, all computational processes converge component-wise to the solution in the Hilbert space norm, except for the control, which converges to the solution in the weak topology, which corresponds to the nature of the phenomenon.

We emphasize that only evidence-based optimization transforms mathematical models into a tool for obtaining guaranteed solutions. Moreover, evidence-based optimization is a beautiful extension of the idea of Lyapunov stability.

References

1. Alekseev, V.M., Tikhomirov, V.M., Fomin, S.V.: Optimal Control. Fizmatlit, Moscow (2018) (1979)
2. Antipin, A.S., Khoroshilova, E.V.: Linear programming and dynamics. Ural Math. J. 1(1), 3–19 (2015)
3. Antipin, A.S., Khoroshilova, E.V.: Optimal control with connected initial and terminal conditions. Proc. Steklov Inst. Math. 289(1), 9–25 (2015). https://doi.org/10.1134/S0081543815050028
4. Antipin, A.S., Khoroshilova, E.V.: Saddle-point approach to solving problem of optimal control with fixed ends. J. Global Optim. 65(1), 3–17 (2016)
5. Antipin, A., Khoroshilova, E.: On methods of terminal control with boundary-value problems: Lagrange approach. In: Goldengorin, B. (ed.) Optimization and Its Applications in Control and Data Sciences. SOIA, vol. 115, pp. 17–49. Springer, Cham (2016). https://doi.org/10.1007/978-3-319-42056-1_2
6. Antipin, A.S., Khoroshilova, E.V.: Lagrangian as a tool for solving linear optimal control problems with state constraints. In: Proceedings of the International Conference on Optimal Control and Differential Games Dedicated to L.S. Pontryagin on the Occasion of His 110th Birthday, pp. 23–26 (2018)
7. Antipin, A.S., Khoroshilova, E.V.: Controlled dynamic model with boundary-value problem of minimizing a sensitivity function. Optim. Lett. 13(3), 451–473 (2019)
8. Antipin, A.S., Khoroshilova, E.V.: Dynamics, phase constraints, and linear programming. Comput. Math. Math. Phys. 60(2), 184–202 (2020)
9. Antipin, A., Khoroshilova, E.: Saddle-point method in terminal control with sections in phase constraints. In: Olenev, N., Evtushenko, Y., Khachay, M., Malkova, V. (eds.) OPTIMA 2020. LNCS, vol. 12422, pp. 17–26. Springer, Cham (2020). https://doi.org/10.1007/978-3-030-62867-3_2
10. Antipin, A., Khoroshilova, E.: Optimal control of two linear programming problems. In: Olenev, N.N., Evtushenko, Y.G., Jaćimović, M., Khachay, M., Malkova, V. (eds.) OPTIMA 2021. LNCS, vol. 13078, pp. 151–164. Springer, Cham (2021). https://doi.org/10.1007/978-3-030-91059-4_11
11. Antipin, A.S., Jaćimović, V., Jaćimović, M.: Dynamics and variational inequalities. Comput. Math. Math. Phys. 57(5), 784–801 (2017). https://doi.org/10.1134/S0965542517050013
12. Antipin, A.S., Vasilieva, O.O.: Dynamic method of multipliers in terminal control. Comput. Math. Math. Phys. 55(5), 766–787 (2015). https://doi.org/10.1134/S096554251505005X

13. Rao, A.V.: A survey of numerical methods for optimal control. Adv. Astronaut. Sci. **135**(1) 09–334 (2010)
14. Vasilyev, F.P.: Optimization methods. In 2 Books. Moscow Center for Continuous Mathematical Education (2011)
15. Vasiliev, F.P., Khoroshilova, E.V., Antipin, A.S.: Regularized Extragradient Method for Finding a Saddle Point in an Optimal Control Problem. Proc. Inst. Math. Mech. Ural Branch Russian Acad. Sci. **17**(1), 27–37 (2011)

Qualitative Analysis of an Infinite Horizon Optimal Control Problem of a Shallow Lake

Dmitry Gromov[1]([✉])([iD]) and Yilun Wu[2]

[1] Department of Mathematics, University of Latvia, Raiņa bulvāris 19, Rīga 1586, Latvia
dv.gromov@gmail.com
[2] Faculty of Applied Mathematics and Control Processes, St. Petersburg State University, 199034 St. Petersburg, Russia

Abstract. This paper studies a classical infinite horizon optimal control problem for a shallow lake model and a variation thereof. We carry out a qualitative analysis of solutions to the canonical system and identify possible scenarios. Specifically, we describe a particular case that has not been addressed in the previous works. This case corresponds to the situation, when the canonical system has only two saddle equilibrium points without a source between them. Furthermore, the set of parameters, for which this situation occurs remains unchanged for two alternative formulations of the optimal control problem, which indicates a possibility for a hidden invariant structure. Both formulations of the optimal control problem are studied in detail, both analytically and numerically. The appearance of the Skiba point is discussed.

Keywords: Shallow lake · Optimal control · Infinite horizon · Skiba point

1 Introduction

Among many mathematical models describing the complex character of processes arising in ecological systems, the shallow lake model plays an important role, cf. [18]. With its non-linear dynamics characterized by multiple steady states and multiple domains of attraction, this model is best suited for describing the evolution of ecosystems featuring both resilience within a domain of stability and an abrupt regime shift outside it. While the dynamics of a shallow lake is interesting on its own, many new phenomena arise when we consider the problem of optimal management and collective use of such a system. The optimal control problem for a shallow lake was first formulated by Mäler, Xepapadeas, and de Zeeuw in [12] and further developed along different directions in a series of papers by different

---

A part of this study was carried out while the first author was with the Faculty of Applied Mathematics and Control Processes, St. Petersburg State University.

authors, see, e.g., [3, 8, 9] to mention just a few. For a detailed overview of the results devoted to the shallow lake model, see Gromov and Upmann [7].

One particular direction of study consists in analyzing the qualitative properties of the optimal solutions to the discounted optimal control problem of special form, considered over the infinite horizon. The main results of this study were presented by Wagener in [17] and further refined by Kiseleva and Wagener in [10]. This research program is concentrated around studying the equilibria of the respective canonical system and analyzing the qualitative properties and performing sensitivity and bifurcation analysis of the optimal solutions, which are identified as the stable manifolds corresponding to the saddle-type equilibrium points.

A particularly interesting case occurs when there are 3 equilibrium points, 2 of which are of the saddle point-type. In this case, there are two candidates to optimal solution, and the choice of the optimal one depends on the initial value of the state variable and the values of the system parameters. Furthermore, it was observed in [17] that as the values of parameters vary, the system undergoes heteroclinic bifurcations that are accompanied by an abrupt change in the structure of optimal solutions (see [11] for more details on bifurcation theory and [1, 6, 14] for several examples of the application of saddle point bifurcations in different fields).

In this paper, we analyze the described problem from a slightly different perspective and report certain features that seem not to have been addressed so far. First, we carry out the parametric analysis in terms of three parameters instead of only two. As we will argue below, this is because one cannot fix the discount rate to a specific value, since after performing non-dimensionalization the new, dimensionless discount rate is not related to the real, physical time any longer. Further, in contrast to the works [10, 17], we consider the optimal control problem with respect to the state and the adjoint variables (instead of the state vs. control analysis) and identify the situations when one or more equilibrium points are unfeasible due to the restrictions imposed on the control variable (which, in turn, depends on the adjoint variable). Specifically, we have to exclude the equilibrium points satisfying $\psi^* > 0$, as this would lead to negative values of the optimal control. In this way, we identify and describe in detail the case of 2 equilibrium points and derive the condition on the values of parameters that corresponds to this case.

Furthermore, we consider two alternative formulations of the cost function: the linear and the quadratic one. This choice is due to the fact that the quadratic cost function may not always adequately describe the observed dependencies, and it is thus of interest to consider the alternative scenarios. It is shown that while these two cases demonstrate qualitatively the same picture, the shape of the regions in the parameter space corresponding to different cases vary to some extent. However, it also appears that the conditions for the scenario with 2 equilibria to realize remain invariant for both cases. This is a remarkable fact that deserves further investigation. Finally, we carried out numerical computations for both types of the cost function and shown the appearance of the Skiba points.

The rest of this paper is organized as follows. In Sect. 2, we introduce the shallow lake model and define two optimal control problems to be considered.

The qualitative analysis of the formulated optimal control problems is carried out in Sect. 3. All obtained results are illustrated with numerical simulations. The last section contains the concluding remarks.

2   Problem Statement

2.1   The Shallow Lake Model

The shallow lake model, first introduced in [12], describes the dynamics of the stock of phosphorus accumulated in the water and the algae, which is denoted by $P(t)$:

$$\dot{P}(\tau) = I(\tau) - (s + h)P(\tau) + f(P(\tau)). \tag{1a}$$

Here, $I(\tau)$ is the non-negative inflow of phosphorus due to agricultural activity (*the loading rate*), s is the sedimentation rate, h is the hydrologic loss rate, and the function f describes the recuperation of phosphorus from the sediments, referred to as the *internal loading*. This is a convex-concave function, described by the following equation

$$f(P) = r\frac{P^2(\tau)}{P^2(\tau) + m^2}, \tag{1b}$$

where $r > 0$ is the maximum rate of internal loading, while $m > 0$ is the shape coefficient. The initial amount of phosphorus is equal to $P(0) = P_0 > 0$.

By an appropriate change of variables, $x = \dfrac{P}{m}$, $u = \dfrac{I}{r}$, $b = \dfrac{(s+h)m}{r}$, and rescaling the time variable as $\tau = \dfrac{r}{m}t$, Eq. (1) is non-dimensionalized and writes as:

$$\dot{x}(t) = u(t) - bx(t) + \frac{x^2(t)}{x^2(t) + 1}. \tag{2}$$

The differential Eq. (2) has either 1 or 3 positive equilibrium points and for sufficiently small values of b and slowly varying u it exhibits two saddle-node bifurcations, which result in a hysteresis-like behavior of a shallow lake (see [7] for a detailed exposition).

2.2   The Optimal Control Problem

In the literature on the optimal control of shallow lakes, it is common to consider the instantaneous profit function as a difference between the logarithmic utility function $\ln(u)$ and the quadratic cost function cx^2 (see, e.g., [12,17]). While this choice is well motivated, under some circumstances different utility, resp. cost functions may better reflect the observed dependencies. In what follows, we consider two variants of the cost function, i.e., a linear cost function and a quadratic one.

We thus consider the following infinite horizon optimization problem:

$$J = \int_0^\infty e^{-\rho t} \left[\ln(u) - \sigma(x) \right] dt \to \max \tag{3a}$$

$$\text{s.t. } \dot{x} = u - bx + \beta(x), \quad x(0) = x_0 > 0, \tag{3b}$$

where $\rho > 0$ is the (dimensionless) depreciation rate, $\beta(x) = \dfrac{x^2}{1 + x^2}$ is the normalized internal loading function, and $\sigma(x)$ is the cost function, which is assumed to be either linear, $\sigma(x) = cx$, or quadratic, $\sigma(x) = cx^2$. Note, however, that the coefficient c is technically not the same in both cases, as it corresponds to two different modelling assumptions about the cost function.

Before proceeding to the subsequent analysis, we wish to stress an important fact. The optimal control problem (3) is formulated for a non-dimensionalized system. This implies that the time variable t does not coincide with the physical time τ any longer and hence, the depreciation rate ρ cannot be chosen to be a sufficiently small number within the range $(0.01, 0.05)$ as it is common in applications. Instead, the parameter ρ can vary within a large range. To illustrate that, let γ be the "real" depreciation rate. Then the non-dimensionalized depreciation rate is given by $\rho = \gamma \dfrac{r}{m}$, which may deviate from γ by orders of magnitude depending on the value of the fraction $\dfrac{r}{m}$. Thus, the parameter ρ is assumed to take an arbitrary value in the subsequent analysis.

To solve the optimal control problem (3) we use the Pontryagin Maximum Principle (see [13] for the original treatment and [5] for a modern exposition). The current value of the Hamiltonian function is

$$H(x, \psi, u) = \ln(u) - \sigma(x) + \psi(u - bx + \beta(x)), \tag{4}$$

whence we have that the optimal control is $u^* = -1/\psi$, which, along with the lower bound on u, results in

$$u^* = \begin{cases} -\frac{1}{\psi}, & \psi < 0; \\ 0, & \psi > 0. \end{cases}$$

In the case $\psi = 0$, there is an uncertainty, as the optimal control takes either a positive or negative infinite value depending on how the adjoint variable approaches 0. We will consider this case separately.

The differential equation for the current value of the adjoint variable ψ is

$$\dot{\psi} = \rho\psi - \frac{\partial H}{\partial x}.$$

So, the canonical system is

$$\dot{x} = u^* - bx + \beta(x) \tag{5a}$$

$$\dot{\psi} = \rho\psi + \sigma'(x) + b\psi - \psi\beta'(x). \tag{5b}$$

Although the canonical system (5) has highly non-linear dynamics, one can readily observe that the trace of the Jacobian matrix of (5) is constant and equals ρ for all admissible values of x and ψ. This implies that any equilibrium point of the system (5) is either a source or a saddle point. This implies, in particular, that the only bounded solution to (5) is the one that corresponds to the stable branch of the saddle point. Such trajectories are the natural candidates for the optimal solution.

3   Qualitative Analysis

3.1   Linear Cost Function $\sigma(x) = cx$

In what follows, we will present a qualitative analysis of the system (5) for the case $\sigma(x) = cx$. We consider the region $\mathcal{R}_- = \{(x, \psi) \in \mathbb{R}^2 | x \geq 0, \psi < 0\}$, which corresponds to the positive finite values of the optimal control u^*. The respective canonical system thus writes as

$$\dot{x} = -\frac{1}{\psi} + \beta(x) \qquad\qquad = -\frac{1}{\psi} - bx + \frac{x^2}{1 + x^2} \tag{6a}$$

$$\dot{\psi} = \rho\psi + \sigma'(x) + b\psi - \beta'(x)\psi \qquad = \rho\psi + c + b\psi - \frac{2x}{(1 + x^2)^2}\psi. \tag{6b}$$

System's Behavior on the Boundary. Consider the behavior of the system on the boundaries of the region $\mathcal{R}_-$. Let $x = 0$. This corresponds to $\dot{x} > 0$ for all $(0, \psi) \in \mathcal{R}_-$, i.e., the vector field of (6) is directed toward the interior of $\mathcal{R}_-$ for $x = 0$. If, on the other hand, we let $\psi \nearrow 0^-$, the respective vector field approaches a horizontal direction: $(\dot{x}, \dot{\psi}) \rightarrow (\infty, c)$. This implies that the trajectories starting in the close vicinity of the boundary of $\mathcal{R}_-$ may leave this region only asymptotically, as x goes to infinity. We do not explore this option any further, but rather mention that a numerical analysis indicates that the region $\mathcal{R}_-$ is invariant under (6).

Equilibrium Points. To determine the number and the type of the equilibrium points, we set the right-hand sides of (6) to zero and eliminate ψ from the resulting system of equations. This yields a rational fraction, whose denominator takes zero values for $x \in \left\{0, \frac{1}{2b}\left(1 - \sqrt{1 - 4b^2}\right), \frac{1}{2b}\left(1 + \sqrt{1 - 4b^2}\right)\right\}$. None of these poles corresponds to an equilibrium point. Thus, we consider the numerator of the fraction, which is given by the following fifth-order polynomial in x:

$$p_1(x) = bcx^5 - (b + c + \rho)\,x^4 + (2bc)\,x^3 - (2b + c + 2\rho)\,x^2 + (bc + 2)\,x - (b + \rho). \tag{7}$$

Using Descartes' rule of signs, we conclude that this polynomial has either 5, 3, or 1 positive real roots. Furthermore, there are no negative real roots, as can be easily concluded from the analysis of $p_1(-x)$. However, a detailed analysis of the nullclines for the system (6) indicates that there can never be 5 equilibrium points, i.e., the polynomial $p_1(x)$ has at most 3 positive real roots.

The positive real roots of $p_1(x)$, taken together with the respective values for ψ obtained from $\psi = \left(-bx + \dfrac{x^2}{1+x^2}\right)^{-1}$, define the candidates for the equilibrium points of the system (6). To determine the proper equilibria, one has to exclude the points that lie outside $\mathcal{R}_-$, that is the points with $\psi > 0$. The situation, when there is a single inadmissible point, can be characterized relatively easily. It corresponds to the case when the polynomial $p_1(x)$ takes the values of different sign at the boundary points of the interval

$$I = \left[\frac{1}{2b}\left(1 - \sqrt{1 - 4b^2}\right), \frac{1}{2b}\left(1 + \sqrt{1 - 4b^2}\right)\right]. \tag{8}$$

This amounts to checking the condition

$$\rho < b\sqrt{1 - 4b^2}, \tag{9}$$

which, quite remarkably, does not depend on the value of c. Figure 1 illustrates the above analysis by presenting the decomposition of the parameters space (b, ρ) for different values of c.

While the qualitative properties of solutions corresponding to the regions of the parameter space that result in either 1 or 3 equilibria are well studied (see, e.g., [17]), the case of 2 equilibrim points does not seem to have been addressed yet. Based upon the conducted numerical analysis, we conclude that for the values of b and ρ that satisfy (9) both equilibrium points are always of the saddle type and hence the optimal solutions coincide with the stable manifolds of these equilibria. The respective stable manifolds are illustrated in Fig. 2a. When extending the stable branches that are located between the equilibrium points backward in time, one can observe that they asymptotically converge to the set of unstable equilibria of (6a). Moreover, at the point where the stable branch of the right saddle crosses the set of these equilibria, the respective trajectory passes a bend. This effectively implies that for a certain range of initial values of the state variable, there appear to be three potential candidates for the optimal solution: two on the same branch and one on the other one.

This observation opens up the question about the existence and location of the so-called *indifference*, or Skiba points [4, 15, 16]. To address this question, we compute the value of the profit functional along the respective stable branches of the saddle-type equilibrium points of (6a) as shown in Fig. 2b. To do so, we recall that the value of the discounted functional of form (3a) is given by [5, Prop. 3.75]

$$\int_0^\infty e^{-\rho t}[\ln(u) - \sigma(x)]dt = \frac{1}{\rho}H(x(0), \psi(0), u(0)),$$

where the initial value of the Hamiltonian function, $H(x(0), \psi(0), u(0))$, is computed numerically along the respective stable branches. Thus, the evaluation of the profit functional boils down to computing the value of the Hamiltonian (4).

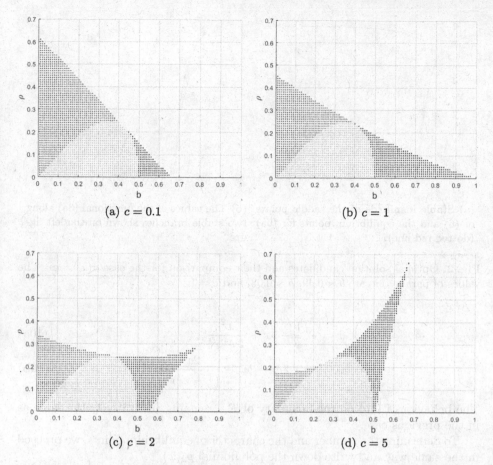

Fig. 1. Linear cost function. Decomposition of the parameter space (b, ρ) for different values of c. The colors indicate the following cases: blue is for 3 equilibria, all admissible; magenta is for 3 equilibria, one of which is inadmissible; red is for 3 equilibria, two of which are inadmissible. The blank region contains the values of (b, ρ) corresponding to a single admissible equilibrium. (Color figure online)

First, we observe that the solution on the bent branch that starts below the bend is clearly not optimal. Further, the intersection point of two curves on Fig. 2b is exactly the Skiba point, i.e., the point that belongs to two optimal solutions.

3.2 Quadratic Cost Function $\sigma(x) = cx^2$

Next, we consider the system (5) for the case $\sigma(x) = cx^2$. The expression for the optimal control remains unchanged. Hence, we again restrict ourselves to the region $\mathcal{R}_-$. The respective canonical system thus writes as follows:

128   D. Gromov and Y. Wu

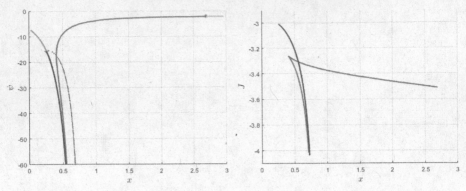

(a) Stable manifolds of the saddle points of (6) and the equilibrium points for (6a) (dotted red line).

(b) The values of the functional (3a) along two stable branches shown on the left figure.

Fig. 2. Optimal solution candidates and their comparison for the case $\sigma(x) = cx$. The values of parameters are $b = 0.49$, $\rho = 0.03$, and $c = 1$.

$$\dot{x} = -\frac{1}{\psi} - bx + \frac{x^2}{1+x^2} \tag{10a}$$

$$\dot{\psi} = \rho\psi + 2cx + b\psi - \frac{2x}{(1+x^2)^2}\psi. \tag{10b}$$

The behavior of (10) on the boundary of $\mathcal{R}_-$ remains qualitatively the same as in the previous case.

To determine the number and the character of equilibrium points, we proceed in the same way and write down the polynomial $p_2(x)$,

$$p_2(x) = 2bcx^6 - 2cx^5 + (4bc - \rho - b)x^4 - 2cx^3 + 2(bc - \rho - b)x^2 + 2x - (b+\rho), \tag{11}$$

whose roots, taken along with the respective values of the adjoint variable, give the candidates to equilibria. Applying Descartes' rule of signs, we conclude that there can be at most 5 positive real roots if $4bc > \rho + b$ and at most 3 positive real root if $4bc < \rho+b$. Furthermore, the analysis of nullclines indicates that the number of equilibrium points cannot exceed 3.

It remains to determine the number of inadmissible equilibrium point, i.e., the equilibrium points that lie outside $\mathcal{R}_-$. Following the same procedure, we check the condition under which $p_2(x)$ has exactly one inadmissible root (note that the case when all 3 real positive roots belong to the interval I (8) can be excluded on the base of the nullcline analysis). Surprisingly, it turns out that the values of parameters corresponding to exactly two equilibria satisfy again the condition (9). To resolve this seeming conundrum. we note that the polynomials (7) and (11) are related as

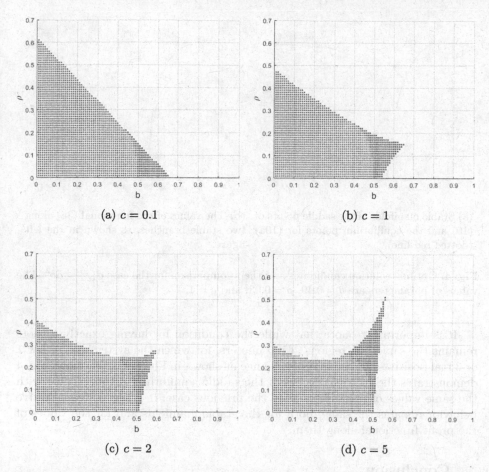

(a) $c = 0.1$

(b) $c = 1$

(c) $c = 2$

(d) $c = 5$

Fig. 3. Quadratic cost function. Decomposition of the parameter space (b, ρ) for different values of c. The colors indicate the following cases: blue is for 3 equilibria, all admissible; magenta is for 3 equilibria, one of which is inadmissible; red is for 3 equilibria, two of which are inadmissible. The blank region contains the values of (b, ρ) corresponding to a single admissible equilibrium. (Color figure online)

$$p_2(x) = p_1(x) + cx(2x - 1)(x^2 + 1)h(x),$$

where $h(x) = bx^2 - x + b$ equals 0 exactly at the boundary points of the interval I. This implies that the resultants (see [2])

$$\mathcal{R}_x(p_1(x), h(x)) = \mathcal{R}_x(p_2(x), h(x)) = 4 + \frac{\rho^2 - b^2}{b^4}$$

are equal. Setting the latter equation to zero, we obtain the expression for the curve separating the region with 2 equilibria.

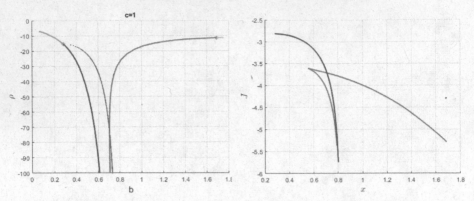

(a) Stable manifolds of the saddle points of (10) and the equilibrium points for (10a) (dotted red line).

(b) The values of the functional (3a) along two stable branches, as shown in the left figure.

Fig. 4. Optimal solution candidates and their comparison for the case $\sigma(x) = cx^2$. The values of parameters are $b = 0.49$, $\rho = 0.03$, and $c = 1$.

It is important to note that while the condition for having exactly 2 roots remains the same, the regions of parameters, for which the system has either 1 or 3 real positive admissible roots, change as shown in Fig. 3. Furthermore, Fig. 4 demonstrates the stable branches of the saddle equilibrium points of (10) with the same values of parameters as in the previous case. It is seen the qualitative behavior remains the same, both for the stable manifolds and for the values of the profit functional along them.

4   Conclusion

In this paper, we studied an infinite horizon optimal control problem for a shallow lake model. Both a classical optimal problem with logarithmic utility and quadratic cost and a variation thereof, in which the cost was assumed to be linear, were considered. We analyzed the equilibrium points of the canonical systems for both optimal control problems and identified a specific case that does not seem to have been addressed in the previous studies. In contrast to the "standard" situation, when there are two saddle point equilibria and a source between them, for certain values of parameters it happens that there are only two saddle point equilibria. In the latter case, for certain initial values of the state, there are three candidates for optimal solutions. To determine the optimal solution, the value of the profit functional was calculated along the stable manifolds of the respective saddle points and the appearance of indifference, or Skiba points was demonstrated. The reported study will be continued by studying qualitative properties of optimal solutions for other classes of profit functionals. This task is particularly important as it would allow putting the obtained results on the qualitative properties of optimal solutions in a wider context.

Acknowledgements. We thank Prof. A. Yu. Uteshev for valuable hints regarding the polynomial analysis. The reported study was funded by RFBR and DFG, project number 21-51-12007.

References

1. Agliari, A., Vachadze, G.: Homoclinic and heteroclinic bifurcations in an overlapping generations model with credit market imperfection. Comput. Econ. **38**(3), 241–260 (2011). https://doi.org/10.1007/s10614-011-9282-y
2. Bikker, P., Uteshev, A.Y.: On the Bézout construction of the resultant. J. Symb. Comput. **28**(1–2), 45–88 (1999). https://doi.org/10.1006/jsco.1999.0267
3. Brock, W.A., Starrett, D.: Managing systems with non-convex positive feedback. Environ. Res. Econ. **26**(4), 575–602 (2003). https://doi.org/10.1023/B:EARE. 0000007350.11040.e3
4. Caulkins, J.P., Feichtinger, G., Grass, D., Hartl, R.F., Kort, P.M., Seidl, A.: Skiba points in free end-time problems. J. Econ. Dyn. Control **51**, 404–419 (2015). https://doi.org/10.1016/j.jedc.2014.11.003
5. Grass, D., Caulkins, J.P., Feichtinger, G., Tragler, G., Behrens, D.A.: Optimal control of nonlinear processes: With Applications in Drugs, Corruption, and Terror. Springer, Heidelberg (2008). https://doi.org/10.1007/978-3-540-77647-5
6. Gromov, D., Castanos, F.: Sensitivity analysis of limit cycles in an alpha Stirling engine: a bifurcation-theory approach. SIAM J. Appl. Dyn. Syst. **19**(3), 1865–1883 (2020). https://doi.org/10.1137/19M1299293
7. Gromov, D., Upmann, T.: Dynamics and economics of shallow lakes: a survey. Sustainability **13**(24) (2021). https://doi.org/10.3390/su132413763
8. Heijnen, P., Wagener, F.O.O.: Avoiding an ecological regime shift is sound economic policy. J. Econ. Dyn. Control **37**(7), 1322–1341 (2013). https://doi.org/10. 1016/j.jedc.2013.03.003
9. Janmaat, J.A.: Fishing in a shallow lake: exploring a classic fishery model in a habitat with shallow lake dynamics. Environ. Res. Econ. **51**(2), 215–239 (2012). https://doi.org/10.1007/s10640-011-9495-5
10. Kiseleva, T., Wagener, F.O.: Bifurcations of optimal vector fields in the shallow lake model. J. Econ. Dyn. Control **34**(5), 825–843 (2010). https://doi.org/10.1016/ j.jedc.2009.11.008
11. Kuznetsov, Y.A.: Elements of Applied Bifurcation Theory. Springer, Heidelberg (1995). https://doi.org/10.1007/978-1-4757-3978-7
12. Mäler, K.G., Xepapadeas, A., De Zeeuw, A.: The economics of shallow lakes. Environ. Res. Econ. **26**(4), 603–624 (2003). https://doi.org/10.1023/B:EARE. 0000007351.99227.42
13. Pontryagin, L.S., Boltyanskii, V.G., Gamkrelidze, R.V., Mishchenko, E.F.: Mathematical Theory of Optimal Processes. CRC Press, Boca Raton (1987)
14. Qi-Chang, Z., Wei, W., Wei-Yi, L.: Heteroclinic bifurcation of strongly nonlinear oscillator. Chin. Phys. Lett. **25**(5), 1905 (2008). https://doi.org/10.1088/0256-307X/25/5/105
15. Sethi, S.P.: Nearest feasible paths in optimal control problems: theory, examples, and counterexamples. J. Optim. Theory Appl. **23**(4), 563–579 (1977). https://doi. org/10.1007/BF00933297
16. Skiba, A.K.: Optimal growth with a convex-concave production function. Econometrica: J. Econometric Soc., 527–539 (1978). https://doi.org/10.2307/1914229

17. Wagener, F.O.O.: Skiba points and heteroclinic bifurcations, with applications to the shallow lake system. J. Econ. Dyn. Control **27**(9), 1533–1561 (2003). https://doi.org/10.1016/S0165-1889(02)00070-2
18. de Zeeuw, A.: Regime shifts in resource management. Ann. Rev. Res. Econ. **6**(1), 85–104 (2014). https://doi.org/10.1146/annurev-resource-100913-012405

Optimization and Data Analysis

Some Adaptive First-Order Methods for Variational Inequalities with Relatively Strongly Monotone Operators and Generalized Smoothness

Seydamet S. Ablaev[1,2] (ID), Alexander A. Titov[2,3](✉) (ID), Fedor S. Stonyakin[1,2] (ID),
Mohammad S. Alkousa[2,3] (ID), and Alexander Gasnikov[2,3,4] (ID)

[1] V. I. Vernadsky Crimean Federal University, Simferopol, Russia
[2] Moscow Institute of Physics and Technology, Moscow, Russia
{a.a.titov,mohammad.alkousa}@phystech.edu
[3] HSE University, Moscow, Russia
[4] Caucasus Mathematical Center, Adyghe State University, Maikop, Russia

Abstract. In this paper, we introduce some adaptive methods for solving variational inequalities with relatively strongly monotone operators. Firstly, we focus on the modification of the recently proposed, in smooth case [18], adaptive numerical method for generalized smooth (with Hölder condition) saddle point problem, which has convergence rate estimates similar to accelerated methods. We provide the motivation for such an approach and obtain theoretical results of the proposed method. Our second focus is the adaptation of widespread recently proposed methods for solving variational inequalities with relatively strongly monotone operators. The key idea in our approach is the refusal of the well-known restart technique, which in some cases causes difficulties in implementing such algorithms for applied problems. Nevertheless, our algorithms show a comparable rate of convergence with respect to algorithms based on the above-mentioned restart technique. Also, we present some numerical experiments, which demonstrate the effectiveness of the proposed methods.

Keywords: Saddle point problem · Hölder continuity · Variational inequality · Restart technique · Strongly convex programming problem

1 Introduction

Non-smooth convex optimization plays a key role in solving the vast majority of modern applied problems [6, 8, 10, 29]. In this paper, we focus on non-smooth saddle point problems and variational inequalities, which, as can be easily shown,

The work of F. Stonyakin and A. Gasnikov was supported by the strategic academic leadership program ¡¡Priority 2030¿¿ (Agreement 075-02-2021-1316, 30.09.2021).

136   S. S. Ablaev et al.

are closely related to each other [20,30]. Such settings of the optimization problems naturally arise when considering problems in machine learning [12,17], data science [20], economic systems [5], optimal transport [11], network equilibrium [14], game theory [25], general equilibrium theory [9,16], etc.

Remind the problem of solving Minty variational inequality. For a given operator $g : X \to \mathbb{R}^n$, where X is a closed convex subset of some finite-dimensional vector space, we need to find a vector $x_* \in X$, such that

$$\langle g(x), x_* - x \rangle \leqslant 0, \quad \forall x \in X. \tag{1}$$

The operator g is called L-smooth, if for any $x, y, z \in X$ the following inequality holds

$$\langle g(y) - g(z), x - z \rangle \leqslant LV(x, z) + LV(z, y), \tag{2}$$

where $V(\cdot, \cdot)$ is the distance in some generalized sense, namely, Bregman divergence (see (9), below). We also assume, that the operator g is μ–relatively strongly monotone, i.e.

$$\mu V(y, x) + \mu V(x, y) \leqslant \langle g(y) - g(x), y - x \rangle, \quad \forall x, y \in X. \tag{3}$$

The concept of relative strong monotonicity is a natural generalization of the concept of relative strong convexity of the objective functional [23] in optimization problems, for variational inequalities.

We are motivated by the following saddle point problem

$$\min_x \max_y f(x, y). \tag{4}$$

Using the recently proposed new paradigm of convex optimization [3,27], namely, relative smoothness condition, there was proposed a technique [18], which provides the possibility of the acceleration of numerical methods for solving saddle point problems, assuming, that the gradient of the objective function $\nabla f(x, y)$ satisfies Lipschitz condition. Moreover, the proposed method is adaptive with respect to all Lipschitz constants of the objective's gradient.

In this paper, we extend the considered class of saddle point problems and replace the classical Lipschitz continuity condition

$$\|\nabla f(z) - \nabla f(u)\| \leqslant L\|z - u\|, \tag{5}$$

by the following Hölder continuity condition

$$\|\nabla f(z) - \nabla f(u)\| \leqslant L_\nu \|z - u\|^\nu, \tag{6}$$

$z = (x_z, y_z), u = (x_u, y_u)$, with respect to the gradient of the objective, where $0 \leqslant \nu \leqslant 1$. Hereinafter we consider arbitrary non-Euclidean norms, which are defined on the corresponding spaces unless otherwise stated.

Note, that the concept of Hölder continuity is an extremely important generalization of the Lipschitz continuity condition. A huge number of applied problems can be formulated exclusively on the class of minimization of Hölder continuous functionals, e.g. smooth multi-armed bandit problem [21], detecting heart rate variability [26], etc. Also, if some function is uniformly convex, then its conjugated will necessarily have the Hölder-continuous gradient, according to [28].

Based on the recently proposed restart technique of the Universal Proximal Method for solving variational inequalities [15], we propose algorithms, which ensure the ε–approximate solution of the considered problem (1) after no more than

$$N = \left\lceil \frac{2L\Omega}{\mu} \log_2 \frac{R_0^2}{\varepsilon} \right\rceil \tag{7}$$

iterations, where μ denotes the constant of relative strong monotonicity of g, Ω and R_0 are specified in Algorithm 2 and can be understood as some characteristics of the domain of the operator g.

The paper consists of an introduction and four main sections. In Sect. 2, we discuss approach [7,18] to the accelerated rates of first-order methods for strongly convex-concave saddle point problems basing on relative smoothness and strong monotonicity conditions. Moreover, we consider generalizations of smoothness conditions for saddle point problems. In Sect. 3, we propose adaptive version of restarted Mirror Prox method [31] for generalized smooth problems. Further, in Sect. 4, we propose some adaptive methods, which do not imply the restart technique, but have the similar convergence rate estimates of the proposed algorithm with restart technique. In Sect. 5, we present some numerical experiments for the saddle point problem and Minty variational inequality, which demonstrate the effectiveness of the proposed methods.

The contributions of the paper can be formulated as follows.

- We consider the non-smooth strongly convex-concave saddle point problem and propose a restarted version of the Universal Proximal Method for the corresponding variational inequality, which guarantees the ε–approximate solution of the problem (1) in an optimal rate of $N = \left\lceil \frac{2L\Omega}{\mu} \log_2 \frac{R_0^2}{\varepsilon} \right\rceil$ iterations. These algorithms are of interest in case of a huge value of the condition number $\frac{L}{\mu}$ as well as in the case of considering not strongly convex-concave saddle point problems.
- We propose methods beyond the restart technique and show, that in some cases they may have even better estimates of the convergence rate compared to methods, based on the restart technique. Moreover, the required number of iterations of such algorithms does not exceed $N = \left\lceil \frac{L+\mu}{\mu} \log_2 \frac{R_0^2}{\varepsilon} \right\rceil$.
- We present some numerical experiments, which demonstrate the effectiveness of the proposed methods.

We start with some auxiliaries. Let E be a finite-dimensional vector space and E^* be its dual. Let us choose some norm $\|\cdot\|$ on E. Define the dual norm $\|\cdot\|_*$ as follows

$$\|\phi\|_* = \max_{\|x\|\leqslant 1} \{\langle \phi, x \rangle\}, \tag{8}$$

where $\langle \phi, x \rangle$ denotes the value of the linear function $\phi \in E^*$ at the point $x \in E$.

Let us choose some prox-function $d(x)$, which is continuously differentiable and convex on E, and define the corresponding Bregman divergence as follows

$$V(y,x) = V_d(y,x) = d(y) - d(x) - \langle \nabla d(x), y - x \rangle, \quad \forall x, y \in E. \tag{9}$$

The Bregman divergence can be understood as some generalization of the distance in the considered set.

2 Towards Adaptive Accelerated Rates for Saddle Point Problems with Generalized Smoothness Condition

Let $Q_x \subset \mathbb{R}^n$ and $Q_y \subset \mathbb{R}^m$ be nonempty, convex and compact sets. Consider the following saddle point problem

$$\min_{x \in Q_x} \max_{y \in Q_y} \{f(x,y) + h(x) - g(x)\}, \tag{10}$$

where $f(x,y) : Q_x \times Q_y \to \mathbb{R}$ is convex function for fixed $y \in Q_y$ and concave for fixed $x \in Q_x$, functions $h(x)$ and $g(y)$ are convex on Q_x and Q_y, respectively, and for each $x, x' \in Q_x, y, y' \in Q_y$, we have

$$\|\nabla h(x) - \nabla h(x')\|_* \leqslant L_x \|x - x'\|, \quad \|\nabla g(y) - \nabla g(y')\|_* \leqslant L_y \|y - y'\|,$$

for some $L_x > 0, L_y > 0$.

Remark 1. If $f(x,y) : Q_x \times Q_y \to \mathbb{R}$ is strongly convex function for fixed $y \in Q_y$ and strongly concave for fixed $x \in Q_x$, we can consider the problem (10) with $h(x) = \frac{\|x\|^2}{2}$ and $g(y) = \frac{\|y\|^2}{2}$.

Let us consider the following setting of the problem (10). Suppose, for any $x, x' \in Q_x, \ y, y' \in Q_y, \ L_{xx} > 0, L_{yy} > 0, L_{xy} > 0$, and for some $\nu \in [0,1]$, the following inequalities hold

$$\|\nabla_x f(x,y) - \nabla_x f(x',y')\|_* \leqslant L_{xx}\|x - x'\|^\nu + L_{xy}\|y - y'\|^\nu, \tag{11}$$

$$\|\nabla_y f(x,y) - \nabla_y f(x',y')\|_* \leqslant L_{xy}\|x - x'\|^\nu + L_{yy}\|y - y'\|^\nu. \tag{12}$$

Let ω_x, ω_y be some 1-strongly convex function w.r.t. $\|\cdot\|_{Q_x}, \|\cdot\|_{Q_y}$ respectively, $\alpha : Q_x^* \to \mathbb{R}$ be the convex conjugate of $\frac{\|x\|^2}{2} - \mu_x \omega_x$, and $\beta : Q_y^* \to \mathbb{R}$ be the convex conjugate of $\frac{\|y\|^2}{2} - \mu_y \omega_y$. Then we can consider the following saddle point problem

$$\min_{\substack{x \in Q_x, y \in Q_y, \\ b \in Q_y^*, a \in Q_x^*}} \max \left\{ \langle a, x \rangle + \langle b, y \rangle + \mu_x \omega_x(x) - \mu_y \omega_y(y) + f(x,y) - \alpha(a) + \beta(b) \right\}. \tag{13}$$

It is shown [18], that if (x^*, y^*, a^*, b^*) is the saddle point to (13), then (x^*, y^*) is the saddle point to (10). Also, α is $\frac{1}{L_x}$-strongly convex in $\|\cdot\|_{Q_x,*}$, and β is $\frac{1}{L_y}$-strongly convex in $\|\cdot\|_{Q_y,*}$ [18].

Lemma 1. *Define the following operator* $((x, y) \in Q_x \times Q_y := X)$

$$g(x, y, a, b) = \Big(a + \mu_x \nabla \omega_x(x) + \nabla_x f(x, y), -b + \mu_y \nabla \omega_y(y) - \nabla_y f(x, y),$$
$$- x + \nabla \alpha(a), y + \nabla \beta(b)\Big),$$

where f *satisfies* (11)–(12). *Consider the following prox-function*

$$d(x, y, a, b) = \mu_x \omega_x(x) + \mu_y \omega_y(y) + \alpha(a) + \beta(b),$$

and the corresponding Bregman divergence, defined according to (9). *Then* g *is* 1-*relatively strongly monotone, i.e.*

$$\langle g(y) - g(x), y - x \rangle \geqslant V(y, x) + V(x, y), \tag{14}$$

and generalized relatively smooth operator, i.e.

$$\langle g(y) - g(z), y - x \rangle \leqslant LV(y, z) + LV(x, y) + \delta, \tag{15}$$

for some $\delta > 0$, *with*

$$L = \widetilde{L}(\delta) = \left(\frac{2}{\delta}\right)^{\frac{1-\nu}{1+\nu}} \left(\frac{L_{xx}^{\frac{2}{1+\nu}}}{\mu_x} + \frac{L_{xy}^{\frac{2}{1+\nu}}}{\sqrt{\mu_x \mu_y}} + \frac{L_{yy}^{\frac{2}{1+\nu}}}{\mu_y}\right). \tag{16}$$

Proof. The proof is given in arXiv preprint [1].

Hence, considered variational inequalities with relatively strongly monotone and generalized relatively smooth operators allow one to obtain first-order method complexity estimates for the corresponding class of strongly convex-convex saddle point problems, which are similar to the accelerated methods [18]. Moreover, using the artificial inaccuracy, Lemma 1 extends this approach to saddle point problems with generalized smoothness conditions [30,31].

However, extending the class of problems, one can potentially encounter the problem of a large value of $\widetilde{L}(\delta)$. On the other hand, even while considering the smooth case for saddle point problems, it may be difficult to estimate all the 5 parameters μ_x, μ_y, L_{xx}, L_{xy} and L_{yy}. Motivated by this and starting from the methodology of Y. E. Nesterov's works [13,24,28], we propose methods allowing the adaptively selection of the corresponding values of these parameters.

3 Adaptive Restarted Mirror Prox for Variational Inequalities with Relative Strongly Monotone Operators

Recently [31], there was proposed an adaptive universal algorithm (listed as Algorithm 1, below), which can automatically adjust to the smoothness level of the operator g.

Algorithm 1. Universal Mirror Prox for Variational Inequalities [31].

Require: $\varepsilon > 0$, $\delta > 0$, $x_0 \in X$, initial guess $L_0 > 0$, prox-setup: $d(x)$, $V(x, z)$.

1: Set $k = 0$, $z_0 = \arg\min_{u \in X} d(u)$.

2: **repeat**

3:     Find the smaller $i_k \geqslant 0$, such that

$$\langle g(z_k), z_{k+1} - z_k \rangle \leqslant \langle g(w_k), z_{k+1} - w_k \rangle + \langle g(z_k), w_k - z_k \rangle +$$
$$+ L_{k+1}(V(w_k, z_k) + V(z_{k+1}, w_k)) + \delta,$$

    where $L_{k+1} = 2^{i_k - 1} L_k$, and

$$w_k = \arg\min_{x \in X}\{\langle g(z_k), x - z_k \rangle + L_{k+1}V(x, z_k)\},$$

$$z_{k+1} = \arg\min_{x \in X}\{\langle g(w_k), x - w_k \rangle + L_{k+1}V(x, z_k)\}.$$

4: **until** $S_N := \sum\limits_{k=0}^{N-1} \frac{1}{L_{k+1}} \geqslant \frac{\max_{x \in X} V(x, x_0)}{\varepsilon}$.

Ensure: z_N.

Theorem 1 [31]. *Let g be a monotone operator, z_N be the output of Algorithm 1 after N iterations. Then the following inequality holds*

$$\langle g(x_*), z_N - x_* \rangle \leqslant -\frac{1}{S_N}\sum_{k=0}^{N-1}\frac{\langle g(w_k), x_* - w_k \rangle}{L_{k+1}} \leqslant \frac{2LV(x_*, z_0)}{N}. \tag{17}$$

Moreover, the total number of iterations does not exceed

$$N = \left\lceil \frac{2L}{\varepsilon} \cdot \max_{x \in X} V(x_0, x) \right\rceil. \tag{18}$$

Lemma 2. *Let g be a relatively strongly monotone operator. For Algorithm 1, the following δ-decreasing of Bregman divergence takes place*

$$V(x_*, z_N) \leqslant V(x_*, z_0) + \delta S_N. \tag{19}$$

Proof. The proof is given in arXiv preprint [1].

The following Algorithm 2 provides the possibility of the acceleration of the proposed Algorithm 1 for solving variational inequality with relatively strongly monotone operator.

Theorem 2. *Let g be a generalized relatively smooth (15) and μ-relatively strongly monotone operator. Then for the output point x_p of the Algorithm 2, it will be hold: $V(x_*, x_p) \leqslant \varepsilon + \frac{2\Omega L\delta}{\mu^2}$. Moreover, the total number of iterations of Algorithm 2 does not exceed*

$$N = \left\lceil \frac{2L\Omega}{\mu} \cdot \log_2 \frac{R_0^2}{\varepsilon} \right\rceil. \tag{20}$$

Algorithm 2. Restarted version of Algorithm 1.

Require: $\varepsilon > 0$, $\mu > 0$, $\Omega : d(x) \leqslant \frac{\Omega}{2} \; \forall x \in X : \|x\| \leqslant 1$; x_0, R_0 : $V(x_*, x_0) \leqslant R_0^2$.

1: $p = 0, d_0(x) = R_0^2 d\left(\frac{x - x_0}{R_0}\right)$.

2: **repeat**

3:     x_{p+1} — output of Algorithm 1 with prox function $d_p(\cdot)$ and stopping criterion $S_N := \sum_{i=0}^{N-1} L_{i+1}^{-1} \geqslant \frac{\Omega}{\mu}$.

4:     $R_{p+1}^2 = \frac{\Omega R_0^2}{2^{(p+1)} \mu S_{N_p}}$.

5:     $d_{p+1}(x) \leftarrow R_{p+1}^2 d\left(\frac{x - x_{p+1}}{R_{p+1}}\right)$.

6:     $p = p + 1$.

7: **until** $p > \log_2\left(\frac{2R_0^2}{\varepsilon}\right)$.

Ensure: x_p.

Proof. The proof is given in arXiv preprint [1].

Remark 2. As shown above, Algorithm 2 needs no more than $N = \left\lceil \frac{2L\Omega}{\mu} \cdot \log_2 \frac{R_0^2}{\varepsilon} \right\rceil$ iterations to provide a solution of the problem (10), L is defined according to (16), while the technique, described in [33] has the following worse complexity estimate

$$\inf_{\nu \in [0,1]} \left\lceil \left(\frac{L_\nu}{\mu}\right)^{\frac{2}{1+\nu}} \cdot \frac{2^{\frac{2}{1+\nu}} \Omega}{\varepsilon^{\frac{1-\nu}{1+\nu}}} \cdot \log_2 \frac{2R_0^2}{\varepsilon} \right\rceil, \tag{21}$$

where $L_\nu = \tilde{L}\left(\frac{\tilde{L}}{2\varepsilon} \frac{(1-\nu)(2-\nu)}{2-\nu}\right)^{\frac{(1-\nu)(1+\nu)}{2-\nu}}, \tilde{L} = \left(L_{xy}\left(\frac{2L_{xy}}{\mu_y}\right)^{\frac{\nu}{2-\nu}} + L_{xx} D^{\frac{\nu-\nu^2}{2-\nu}}\right)$, and D is the diameter of the domain of $f(x, \cdot)$.

Remark 3. The obtained estimate

$$N = \left\lceil 2\Omega \left(\frac{L_{xx}}{\mu_x} + \frac{L_{xy}}{\sqrt{\mu_x \mu_y}} + \frac{L_{yy}}{\mu_y}\right) \cdot \log_2 \frac{R_0^2}{\varepsilon} \right\rceil \tag{22}$$

is optimal for saddle point problems (13) with $\nu = 1$.

Remark 4. Note, that Ω may depend on the dimension of the considered space [4].

4   First-order Methods for Relatively Strongly Monotone Variational Inequalities Beyond the Restart Technique

Basing on some recently proposed methods [7,18] for VIs with strongly relatively monotone operators $g : X \rightarrow \mathbb{R}^n$, we consider algorithms (see Algorithms 3, 4 and 5) without using the restart technique. Similarly to the previous section we consider the case of operators g with the generalized smoothness condition (15). We improve the quality of the solution, compared to Algorithm 3 by reducing

Algorithm 3. Adaptive first-order method for variational inequalities with μ-relatively strongly monotone and Lipschitz continuous operators without restarts.

Require: $\varepsilon > 0, \delta > 0, x_0 \in X, L_0 > 0, \mu > 0, d(x), V(x, z)$.

1: Set $z_0 = \arg\min_{u \in X} d(u)$.

2: **for** $k \geqslant 0$ **do**

3:    Find the smallest integer $i_k \geqslant 0$, such that

$$\langle g(z_k) - g(w_k), z_{k+1} - w_k \rangle \leqslant L_{k+1}\left(V(w_k, z_k) + V(z_{k+1}, w_k)\right) + \delta, \qquad (23)$$

where $L_{k+1} = 2^{i_k - 1} L_k$, and

$$w_k = \arg\min_{y \in X}\left\{\left\langle \frac{1}{L_{k+1}}g(z_k), y \right\rangle + V(y, z_k)\right\}, \qquad (24)$$

$$z_{k+1} = \arg\min_{z \in X}\left\{\left\langle \frac{1}{L_{k+1}}g(w_k), z \right\rangle + V(z, z_k) + \frac{\mu}{L_{k+1}}V(z, w_k)\right\}. \qquad (25)$$

4: **end for**

Ensure: z_k.

$O\left(\frac{\delta}{\mu^2}\right)$ to $O\left(\frac{\delta}{\mu}\right)$, which provides the better convergence rate in case of small value of μ. It is also worth noting, that proposed algorithms do not require knowledge of the parameter Ω.

Theorem 3. *Let g be a μ-relatively strongly monotone operator, and z_* be the exact solution of the variational inequality (1). Then for Algorithm 3, the following inequality holds*

$$V(z_*, z_{k+1}) \leqslant \prod_{i=1}^{k+1}\left(1 + \frac{\mu}{L_i}\right)^{-1} V(z_*, z_0) + \frac{\delta}{L_{k+1} + \mu}$$

$$+ \sum_{j=1}^{k} \frac{\delta}{L_j + \mu} \prod_{i=j+1}^{k+1}\left(1 + \frac{\mu}{L_i}\right)^{-1}. \qquad (26)$$

Proof. The proof is given in arXiv preprint [1].

Corollary 1. *Let g be a μ-relatively strongly monotone operator, and z_* be the exact solution of the variational inequality (1). Then for Algorithm 4, the following inequalities hold*

$$V(z_*, z_{k+1}) \leqslant \prod_{i=0}^{k}\left(1 + \frac{\mu}{L_{i+1}}\right)^{-1} V(z_*, z_0), \qquad (30)$$

$$V(z_*, z_{k+1}) \leqslant \left(1 + \frac{\mu}{2L}\right)^{-(k+1)} V(z_*, z_0). \qquad (31)$$

Algorithm 4. Adaptive first-order method for variational inequalities with μ-relatively strongly monotone and Lipschitz continuous operator.

Require: $\varepsilon > 0, x_0 \in X, L_0 > 0, \mu > 0, d(x), V(x, z)$.

1: Set $z_0 = \arg\min_{u \in X} d(u)$.

2: **for** $k \geq 0$ **do**

3:    Find smallest integer $i_k \geq 0$, such that

$$\langle g(z_k) - g(w_k), z_{k+1} - w_k \rangle \leq L_{k+1} \left(V(w_k, z_k) + V(z_{k+1}, w_k) \right), \quad (27)$$

where $L_{k+1} = 2^{i_k - 1} L_k$, and

$$w_k = \arg\min_{y \in X} \left\{ \left\langle \frac{1}{L_{k+1}} g(z_k), y \right\rangle + V(y, z_k) \right\}, \quad (28)$$

$$z_{k+1} = \arg\min_{z \in X} \left\{ \left\langle \frac{1}{L_{k+1}} g(w_k), z \right\rangle + V(z, z_k) + \frac{\mu}{L_{k+1}} V(z, w_k) \right\}. \quad (29)$$

4: **end for**

Ensure: z_k.

Remark 5. Due to (31), the number of iterations of Algorithm 4 for solving the problem (1) does not exceed $N = \left\lceil \frac{2L + \mu}{\mu} \log_2 \frac{R_0^2}{\varepsilon} \right\rceil$, which coincides with (20) up to the multiplication by a constant in the case of $L \sim \mu \sim 1$.

Remark 6. Taking into account (16), we find that the inequality (31) has the following form

$$V(z_*, z_{k+1}) = \left(\frac{2\widetilde{L}}{2\widetilde{L} + \mu} \right)^{k+1} V(z_*, z_0),$$

where $\widetilde{L}$ is given in (16).

The main difference between Algorithms 3 and 4 and the next Algorithm 5 is a modified exit criterion, which leads to decreasing of the coefficient at δ.

Theorem 4. *Let g be a μ-relatively strongly monotone operator, and z_* be the exact solution of the variational inequality (1). Then for Algorithm 5, we have*

$$V(z_*, z_{k+1}) \leq \prod_{i=1}^{k+1} \left(1 + \frac{\mu}{L_i} \right)^{-1} V(z_*, z_0) + \delta \left(1 + \sum_{j=1}^{k} \prod_{i=j+1}^{k+1} \left(1 + \frac{\mu}{L_i} \right)^{-1} \right), \quad (35)$$

and

$$V(z_*, z_{k+1}) \leq \left(1 + \frac{\mu}{2L} \right)^{-(k+1)} V(z_*, z_0) + \delta \left(1 + \frac{2L}{\mu} \right). \quad (36)$$

Proof. The proof is given in arXiv preprint [1].

Algorithm 5. The third adaptive first-order method for variational inequalities with μ-relatively strongly monotone and Lipschitz continuous operators.

Require: $\varepsilon > 0, \delta > 0, x_0 \in X, L_0 > 0, \mu > 0, d(x), V(x,z)$.
1: Set $z_0 = \arg\min_{u \in X} d(u)$.
2: **for** $k \geqslant 0$ **do**
3:    Find the smallest integer $i_k \geqslant 0$, such that

$$\langle g(z_k) - g(w_k), z_{k+1} - w_k \rangle \leqslant L_{k+1}(V(w_k, z_k) + V(z_{k+1}, w_k)) + L_{k+1}\delta, \quad (32)$$

where $L_{k+1} = 2^{i_k - 1}L_k$, and

$$w_k = \arg\min_{y \in X}\left\{\left\langle \frac{1}{L_{k+1}}g(z_k), y \right\rangle + V(y, z_k)\right\}, \quad (33)$$

$$z_{k+1} = \arg\min_{z \in X}\left\{\left\langle \frac{1}{L_{k+1}}g(w_k), z \right\rangle + V(z, z_k) + \frac{\mu}{L_{k+1}}V(z, w_k)\right\}. \quad (34)$$

4: **end for**
Ensure: z_k.

Remark 7. If we iterate the inequality

$$V(z_*, z_{k+1}) \leqslant \left(1 + \frac{\mu}{L_{k+1}}\right)^{-1}V(z_*, z_k) + \frac{L_{k+1}\delta}{L_{k+1} + \mu},$$

then for the Algorithm 5, the following inequality holds

$$V(z_*, z_{k+1}) \leqslant \prod_{i=1}^{k+1}\left(1 + \frac{\mu}{L_i}\right)^{-1}V(z_*, z_0) + \frac{L_{k+1}\delta}{L_{k+1} + \mu} + \sum_{j=1}^{k}\frac{L_j\delta}{L_j + \mu}\prod_{i=j+1}^{k+1}\left(1 + \frac{\mu}{L_i}\right)^{-1}.$$

Remark 8. For the selected type of saddle point problems (13) with (16) and $\mu = 1$, we find that the inequality

$$V(z_*, z_{k+1}) \leqslant \left(1 + \frac{\mu}{2L}\right)^{-(k+1)}V(z_*, z_0) + \delta\left(1 + \frac{2L}{\mu}\right),$$

has the following form

$$V(z_*, z_{k+1}) \leqslant \left(1 + \frac{1}{2\widetilde{L}}\right)^{-(k+1)}V(z_*, z_0) + \widetilde{L}2^{\frac{2}{1-\nu}}\delta^{\frac{2\nu(1-\nu)}{(1+\nu)^2}} + \delta,$$

where $\widetilde{L}$ is given in (16).

5   Numerical Experiments

5.1   Saddle Point Problem for the Smallest Covering Ball Problem with Functional Constraints

In this subsection, we consider an example of the Lagrange saddle point problem induced by a problem with geometrical nature, namely, an analogue of the well-known *smallest covering ball problem* with functional constraints. This example

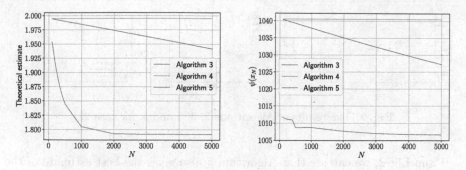

Fig. 1. The results of Algorithms 3, 4 and 5, for case 1.

is equivalent to the following non-smooth convex optimization problem with functional constraints

$$\min_{x \in X} \left\{ \psi(x) := \max_{1 \leqslant k \leqslant s} \|x - A_k\|_2^2; \; \varphi_p(x) \leqslant 0, \; p = 1, ..., m \right\}, \qquad (37)$$

where $A_k \in \mathbb{R}^n, k = 1, ..., s$ are given points and X is a convex compact set. Functional constraints φ_p, for $p = 1, ..., m$, have the following form

$$\varphi_p(x) := \sum_{i=1}^{n} \alpha_{pi} x_i^2 - 5, \quad p = 1, ..., m. \qquad (38)$$

See [1], for more details about the setting of this problem and the setting of its connected conducted experiments with it. The coefficients α_{pi} in (38) are drawn randomly from the following distributions.

Case 1: Pareto II or Lomax distribution with shape equalling 10.

Case 2: chi-square distribution, with a number of degrees of freedom, equals 3.

For case 1, the results of the work of Algorithms 3, 4, and 5 are presented in the Fig. 1, below. These results demonstrate the theoretical estimate (26), (30) and (35) for Algorithms 3, 4 and 5, respectively. Also, they demonstrate the value of the objective function in (37) at the output point x_N of the algorithm after performing N iterations.

For case 2, we compare the work of Algorithms 2, 3, 4, and 5. The results are presented in Fig. 2. They demonstrate the theoretical estimate (26), (30) and (35) for Algorithms 3, 4 and 5, respectively. Also, in the comparison with Algorithm 2, these results demonstrate the value of the objective function at the output point x_N of the algorithms after performing N iterations.

From Fig. 1, we can see that the proposed Algorithm 5 provides the best quality solution with respect to the value of the objective function ψ at the output point x_N, although the theoretical estimate of this quality is not very high. Therefore the efficiency of Algorithm 5 is obvious when we look at the value of the objective function at the output point of compared algorithms.

146   S. S. Ablaev et al.

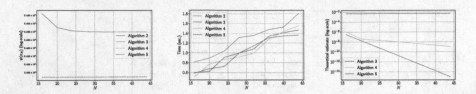

Fig. 2. The results of Algorithms 2, 3, 4 and 5, for case 2.

From Fig. 2, we can see that Algorithm 5 also gives the best estimate of the quality of the solution. Also, Algorithms 3, 4 and 5 give the same objective values at the output points, with approximately the same running time. In this case, we see that Algorithm 4 works better than Algorithm 3 (it gives better estimate of the quality of the solution), not as in case 1.

5.2 One Minty Variational Inequality

In this subsection we consider variational inequality with the following Lipschitz-continuous and strongly monotone operators (see [1], for more details about the setting of this problem and the setting of its connected conducted experiments with it.).

The first operator is (see Example 5.2 in [19])

$$g : X \subset \mathbb{R}^n \to \mathbb{R}^n, \quad g(x) = x. \tag{39}$$

The second operator is

$$g : X \subset \mathbb{R}^n \to \mathbb{R}^n, \quad g(x_1, \dots, x_n) = \left(x_1, 2^2 x_2, 3^2 x_3, \dots, n^2 x_n \right). \tag{40}$$

This operator is L-Lipschitz continuous with $L = n^2$ and μ-strongly monotone with $\mu = 1$. The condition number for this operator is $\kappa = L/\mu = n^2$, therefore this operator will be ill-conditioned when n is relatively big.

For the experiments with operator (39), the results are presented in Fig. 3 and 4, which illustrate the norm $\|x_{\text{out}} - x_*\|_2$, and the running time in seconds as a function of iterations, where x_{out} is the output of each algorithm.

In Fig. 3, we can not see the graphics of the Algorithms 3 and 5, that indicate the distance $\|x_{\text{out}} - x_*\|_2$, because by these algorithms this distance is equal to zero for all considered number of iterations.

From the conducted experiments (for operator (39)), we can see that the shape of the feasible set very much affects the progress of the work of the proposed algorithms. We note that when we increase r (the radius of the ball X), the corresponding running time of the compared algorithms is also increased. Also, from Fig. 3 and 4, we can see that the proposed Algorithms 3 and 5, for any value of the radius r, are the best, where they give the solution of the problem under consideration with very high quality and at the same (approximately) running time. Also, we can see that Algorithm 4 always works better than Algorithm 2.

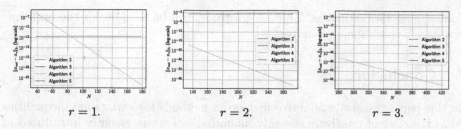

$r = 1.$ $\qquad\qquad$ $r = 2.$ $\qquad\qquad$ $r = 3.$

Fig. 3. The results of Algorithms 2, 4, 3 and 5, for operator (39), in the set $X = \{x \in \mathbb{R}^n, \|x\|_2 \leqslant r\}$ with different radii r and $n = 10^6, \varepsilon \in \{10^{-3i}, i = 1, 2, \ldots, 8\}, \delta = 0.01$.

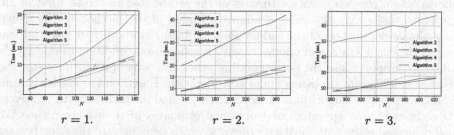

$r = 1.$ $\qquad\qquad$ $r = 2.$ $\qquad\qquad$ $r = 3.$

Fig. 4. The results of Algorithms 2, 4, 3 and 5, for operator (39), in the set $X = \{x \in \mathbb{R}^n, \|x\|_2 \leqslant r\}$ with different radii r and $n = 10^6, \varepsilon \in \{10^{-3i}, i = 1, 2, \ldots, 8\}, \delta = 0.01$.

Now, for the experiments with operator (40), the results are presented in Fig. 5, which illustrates the norm $\|x_{\text{out}} - x_*\|_2$, the running time in seconds as a function of iterations, and the theoretical estimates (19), (26) and (35) of the quality of the solution, for Algorithms 1, 3 and 5, respectively.

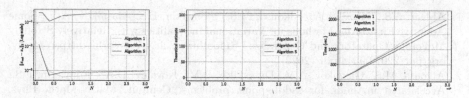

Fig. 5. The results of Algorithms 1, 3 and 5, for operator (40).

From Fig. 5, we can see that Algorithm 5 is the worst. Algorithm 1, with respect to the distance $\|x_{\text{out}} - x_*\|_2$, gives results better than Algorithms 3. Also, the theoretical estimate of the quality of the solution by Algorithm 1 is approximately the same as by Algorithm 3. The difference between the running times of Algorithms 1 and 3 is not big. Note that, Algorithm 1 is applicable to a wider type of problem and can work better than Algorithms 3, 4 and 5 with small μ. At the same time for variational inequalities with strongly monotone operators, it can be restarted in such a way that Algorithm 2 will have similar convergence rate estimates. Thus, Algorithm 2 will be the best for the problems with an ill-conditioned operator.

6   Conclusion

In this paper, we study adaptive first-order methods for variational inequalities from the class of relatively strongly monotone operators recently introduced in [32]. Our research is motivated, in particular, by the recently proposed technique [7,18] for strongly convex-concave saddle point problems, which allows one to obtain the complexity estimates of the accelerated methods. First of all, the paper deals with the issue of adaptive tuning of the method to the global smoothness parameters of the saddle point problem. Moreover, an essential feature is the consideration of operators with a generalized condition of relative L-Lipschitz property and the corresponding generalizations of smoothness for the class of saddle point problems under consideration. Based on the methods from [7,18], we proposed algorithms for solving variational inequalities with relatively strongly monotone operators and obtained estimates of their convergence We also presented some numerical experiments, which demonstrate the effectiveness of the proposed methods. We considered an example of the convex optimization problem with functional constraints, and an example of the Minty variational inequality. The conducted experiments showed that the proposed Algorithms 3, 4 and 5 without using the technique of restarts work better than algorithm with restarts (Algorithm 2) and vice versa for an example of the variational inequality with an ill-conditioned operator.

References

1. Ablaev, S.S., Titov, A.A., Alkousa, M.S., Stonyakin, F.S., Gasnikov, A.V.: Some Adaptive First-order Methods for Variational Inequalities with Relatively Strongly Monotone Operators and Generalized Smoothness. arXiv preprint https://arxiv.org/pdf/2207.09544.pdf (2022)
2. Alkousa, M.S., Gasnikov, A.V., Dvinskikh, D.M., Kovalev, D.A., Stonyakin, F.S.: Accelerated methods for saddle-point problems. Comput. Math. Math. Phys. **60**(11), 1843–1866 (2020)
3. Bauschke, H.H., Bolte, J., Teboulle, M.: A descent lemma beyond Lipschitz gradient continuity: first-order methods revisited and applications. Math. Oper. Res. **42**(2), 330–348 (2017)
4. Ben-Tal, A., Nemirovski, A.: Lectures on modern convex optimization: analysis, algorithms, and engineering applications. Society for industrial and applied mathematics (2001)
5. Buiter, W.H.: Saddle point Problems in Continuous Time Rational Expectations Models: A General Method and Some Macroeconomic Examples. NBER Technical Working Paper No. 20 (1984)
6. Bubeck, S.: Convex optimization: algorithms and complexity. Found. Trends® Mach. Learn. **8**(3–4), 231–357 (2015)

7. Cohen, M. B., Sidford, A., Tian, K.: Relative Lipschitzness in Extragradient Methods and a Direct Recipe for Acceleration. arXiv preprint https://arxiv.org/pdf/2011.06572.pdf (2020)
8. Cheng, L., Hou, Z.G., Lin, Y., Tan, M., Zhang, W.C., Wu, F.X.: Recurrent neural network for non-smooth convex optimization problems with application to the identification of genetic regulatory networks. IEEE Trans. Neural Netw. **22**(5), 714–726 (2011)
9. Cherukuri, A., Gharesifard, B., Cortes, J.: Saddle-point dynamics: conditions for asymptotic stability of saddle points. SIAM J. Control Optim. **55**(1), 486–511 (2017)
10. Clarke, F. H.: Method of Dynamic and Nonsmooth Optimization. Society for Industrial and Applied Mathematics (1989)
11. Dafermos, S.: Traffic equilibrium and variational inequalities. Transp. Sci. **14**(1), 42–54 (1980)
12. Dauphin, Y. N., Pascanu, R., Gulcehre, C., Cho, K., Ganguli, S., Bengio, Y.: Identifying and attacking the saddle point problem in high-dimensional non-convex optimization. Adv. Neural Inf. Process. Syst. **27** (2014)
13. Devolder, O., Glineur, F., Nesterov, Y.: First-order methods of smooth convex optimization with inexact oracle. Math. Program. **146**(1), 37–75 (2014)
14. Friesz, T.L., Bernstein, D., Smith, T.E., Tobin, R.L., Wie, B.W.: A variational inequality formulation of the dynamic network user equilibrium problem. Oper. Res. **41**(1), 179–191 (1993)
15. Gasnikov, A.V., Dvurechensky, P.E., Stonyakin, F.S., Titov, A.A.: An adaptive proximal method for variational inequalities. Comput. Math. Math. Phys. **59**(5), 836–841 (2019)
16. Grandmont, J. M.: Temporary general equilibrium theory. Econometrica: J. Econometric Soc., 535–572 (1977)
17. Jin, C., Netrapalli, P., Ge, R., Kakade, S.M., Jordan, M.I.: On nonconvex optimization for machine learning: gradients, stochasticity, and saddle points. J. ACM (JACM) **68**(2), 1–29 (2021)
18. Jin, Y., Sidford, A., Tian, K.: Sharper Rates for Separable Minimax and Finite Sum Optimization via Primal-Dual Extragradient Methods. arXiv preprint https://arxiv.org/pdf/2202.04640.pdf (2022)
19. Khanh, P.D., Vuong, P.T.: Modified projection method for strongly pseudomonotone variational inequalities. J. Glob. Optim. **58**, 341–350 (2014)
20. Kinderlehrer, D., Stampacchia, G.: An introduction to variational inequalities and their applications. Society for Industrial and Applied Mathematics (2000)
21. Liu, Y., Wang, Y., Singh, A.: Smooth bandit optimization: generalization to holder space. In: International Conference on Artificial Intelligence and Statistics, PMLR, pp. 2206–2214 (2021)
22. Lu, H.: relative continuity for non-lipschitz nonsmooth convex optimization using stochastic (or deterministic) mirror descent. INFORMS J. Optim. **1**(4), 288–303 (2019)
23. Lu, H., Freund, R.M., Nesterov, Y.: Relatively smooth convex optimization by first-order methods, and applications. SIAM J. Optim. **28**(1), 333–354 (2018)
24. Nesterov, Y.: Gradient methods for minimizing composite functions. Math. Program. **140**(1), 125–161 (2013)
25. Mertikopoulos, P., Lecouat, B., Zenati, H., Foo, C. S., Chandrasekhar, V., Piliouras, G.: Optimistic mirror descent in saddle-point problems: Going the extra (gradient) mile. arXiv preprint https://arxiv.org/pdf/1807.02629.pdf (2018)

26. Nakamura, T., Horio, H., Chiba, Y.: Local holder exponent analysis of heart rate variability in preterm infants. IEEE Trans. Biomed. Eng. **53**(1), 83–88 (2005)
27. Nesterov, Y.: Relative smoothness: new paradigm in convex optimization. In Conference report, EUSIPCO-2019, A Coruna, Spain, vol. 4 (2019)
28. Nesterov, Y.: Universal gradient methods for convex optimization problems. Math. Program. **152**(1), 381–404 (2015)
29. Nesterov, Y.: Smooth minimization of non-smooth functions. Math. Program. **103**(1), 127–152 (2005)
30. Stonyakin, F., Gasnikov, A., Dvurechensky, P., Titov, A., Alkousa, M.: Generalized Mirror prox algorithm for monotone variational inequalities: universality and inexact oracle. J. Optim. Theory Appl., 1–26 (2022)
31. Stonyakin, F., et al.: Inexact relative smoothness and strong convexity for optimization and variational inequalities by inexact model. Optim. Methods Softw. **36**(6), 1155–1201 (2021)
32. Stonyakin, F.S., Titov, A.A., Makarenko, D.V., Alkousa, M.S.: Some Methods for Relatively Strongly Monotone Variational Inequalities. arXiv preprint https://arxiv.org/pdf/2109.03314.pdf (2022)
33. Titov, A.A., Stonyakin, F.S., Alkousa, M.S., Gasnikov, A.V.: Algorithms for solving variational inequalities and saddle point problems with some generalizations of lipschitz property for operators. In: Strekalovsky, A., Kochetov, Y., Gruzdeva, T., Orlov, A. (eds.) MOTOR 2021. CCIS, vol. 1476, pp. 86–101. Springer, Cham (2021). https://doi.org/10.1007/978-3-030-86433-0_6
34. Titov, A.A., Stonyakin, F.S., Alkousa, M.S., Ablaev, S.S., Gasnikov, A.V.: Analogues of switching subgradient schemes for relatively lipschitz-continuous convex programming problems. In: Kochetov, Y., Bykadorov, I., Gruzdeva, T. (eds.) MOTOR 2020. CCIS, vol. 1275, pp. 133–149. Springer, Cham (2020). https://doi.org/10.1007/978-3-030-58657-7_13

Compression and Data Similarity: Combination of Two Techniques for Communication-Efficient Solving of Distributed Variational Inequalities

Aleksandr Beznosikov[1,2]([envelope]) and Alexander Gasnikov[1,3,4]

[1] Moscow Institute of Physics and Technology, Dolgoprudny, Russia
anbeznosikov@gmail.com
[2] HSE University, Moscow, Russia
[3] IITP RAS, Moscow, Russia
[4] Caucasus Mathematical Center, Adyghe State University, Maikop, Russia

Abstract. Variational inequalities are an important tool, which includes minimization, saddles, games, fixed-point problems. Modern large-scale and computationally expensive practical applications make distributed methods for solving these problems popular. Meanwhile, most distributed systems have a basic problem – a communication bottleneck. There are various techniques to deal with it. In particular, in this paper we consider a combination of two popular approaches: compression and data similarity. We show that this synergy can be more effective than each of the approaches separately in solving distributed smooth strongly monotone variational inequalities. Experiments confirm the theoretical conclusions.

Keywords: Distributed optimization · Variational inequalities · Compression · Data similarity

1 Introduction

Variational inequalities are a broad class of problems that have been widely studied for a long time. This is primarily due to the uniqueness of variational inequalities; they can describe various types of optimization problems, which, in turn, have many practical applications [7,20]. We can mention classic examples in economics and game theory [19], robust optimization [8], non-smooth optimization [38,40], matrix factorization [6], image denoising [14,18], supervised learning [5]. In recent years, there has been a significant increase in research interest toward the study of variational inequalities due to new connections with GANs [23]. In particular, the authors of [15,17,22,35,36,41] show that even if one considers the classical (in the variational inequalities literature) regime involving

---

The work of A. Beznosikov was supported by the strategic academic leadership program 'Priority 2030' (Agreement 075-02-2021-1316 30.09.2021).

monotone and strongly monotone inequalities, it is possible to obtain insights, methods and recommendations useful for the GANs training.

Until recently, theoretical studies of methods for variational inequalities were carried out only in the non-distributed setting. The `Extra Gradient/Mirror Prox` method [28,31,39] became widely known and very popular. This alorithm for variational inequalities is key and basic (as Gradient Descent for minimization problems). But new practical problems have opened up new challenges. Indeed, the training of modern supervised machine learning models in general, and deep neural networks in particular, becomes more and more demanding. Solving such problems is almost impossible without a distributed approach with parallelization [51].

Meanwhile, the distributed approach has its bottlenecks. The main one is communication cost, as the transfer of information between computing devices takes considerably longer than local processes. This is why it is important not just to get a distributed version of e.g. `Extra Gradient` [12], but a more effective method in terms of communication. The community already knows a number of approaches to communication efficient distributed optimization [21,24,30,46]. For example, two such popular approaches are the compression of transmitted information, and the use of statistical similarity of local data on workers (if we spread the data uniformly among them).

Our Contribution and Related Works. In this work we have combined two techniques for effective communication: compression and data similarity. Through this synthesis, we have obtained a method for distributed variational inequalities with better theoretical guarantees on the number of information transferred. See Table 1.

Table 1. Summary of complexities on the number of transmitted information (bits) for different approaches to communication bottleneck. *Notation:* μ = constant of strong monotonicity of the operator F, L = Lipschitz constant of the operator F, δ = similarity (relatedness) constant, M = number of devices, b = local data size, ε = precision of the solution.

Method	Reference	Technique	Number of bits	If $\delta \sim \frac{L}{\sqrt{b}}$
`Extra Gradient`	[12,28]		$O\left(\frac{L}{\mu}\log\frac{1}{\varepsilon}\right)$	$O\left(\frac{L}{\mu}\log\frac{1}{\varepsilon}\right)$
`SMMDS`	[13]	Similarity	$O\left(\frac{\delta}{\mu}\log\frac{1}{\varepsilon}\right)$	$O\left(\frac{1}{\sqrt{b}}\cdot\frac{L}{\mu}\log\frac{1}{\varepsilon}\right)$
`MASHA`	[11]	Compression	$O\left(\frac{L}{\sqrt{M}\mu}\log\frac{1}{\varepsilon}\right)$	$O\left(\frac{1}{\sqrt{M}}\cdot\frac{L}{\mu}\log\frac{1}{\varepsilon}\right)$
`Optimistic MASHA`	This work	Compression similarity	$O\left(\left[\frac{L}{M\mu}+\frac{\delta}{\sqrt{M}\mu}\right]\log\frac{1}{\varepsilon}\right)$	$O\left(\left[\frac{1}{M}+\frac{1}{\sqrt{Mb}}\right]\cdot\frac{L}{\mu}\log\frac{1}{\varepsilon}\right)$

Separately from each other, similarity and compression techniques have already been investigated for both particular minimization problems and general variational inequalities.

Different approaches with compression have been developed for minimizations. Here we can note the earliest and the simplest approach, in which compression operators were applied to SGD-type methods [3]. Further modifications with "memory" were presented [26,37]. Then accelerated methods were introduced [34]. The work [24] was tried to look at compression through the variance reduction technique. There is now widespread research into practical modifications with biased operators and error compensation [10,29,44,54], bidirectional compression [42,48], partial participation [24,27,42], etc. In the generality of variational inequalities, compression methods were studied in [11]. One can note that our new method are ahead of the MASHA from this paper (record results in terms of compression methods for variational inequalities at the moment).

The literature on distributed minimization problems under similarity (relatedness) assumption is also vast. The paper [4] established lower communication complexity bounds. The authors of [45] proposed a mirror-descent based algorithm with data preconditioning. This technique was further accelerated by the inexact damped Newton method [53], the Catalyst framework [43], the heavy ball momentum [52]. Higher order methods employing preconditioning were studied in [1,16,49]. Not for minimizations, but for variational inequalities and saddles, the similarity (relatedness) setup was considered by [13,32]. In some cases, our estimates can also outperform the results from these works.

2 Problem Setup and Assumptions

2.1 Variational Inequality

We consider variational inequalities (VI) of the form

$$\text{Find}\ \ z^* \in \mathbb{R}^d\ \ \text{such that}\ \ \langle F(z^*), z - z^* \rangle + g(z) - g(z^*) \geq 0,\ \ \forall z \in \mathbb{R}^d, \qquad (1)$$

where $F : \mathbb{R}^d \to \mathbb{R}^d$ is an operator, and $g : \mathbb{R}^d \to \mathbb{R} \cup \{+\infty\}$ is a proper lower semicontinuous convex function. We assume that F is distributed across M workers/devices

$$F(z) := \frac{1}{M} \sum_{m=1}^{M} F_m(z), \qquad (2)$$

where $F_m : \mathbb{R}^d \to \mathbb{R}^d$ for all $m \in \{1, 2, \ldots, M\}$.

2.2 Examples

To showcase the expressive power of the formalism (1), we now give a few examples of variational inequalities arising in machine learning.

Example 1 [Convex minimization]. Consider the composite minimization problem:

$$\min_{z \in \mathbb{R}^d} f(z) + g(z), \qquad (3)$$

where f is typically a main term, and g is a regularizer or an indicator function (e.g. if we want to consider a problem on some set). If we put $F(z) := \nabla f(z)$, then it can be proved that $z^* \in \text{dom}\, g$ is a solution for (1) if and only if $z^* \in \text{dom}\, g$ is a solution for (3).

Example 2 [Convex-concave saddle point problems]. Consider the convex-concave saddle point problem

$$\min_{x \in \mathbb{R}^{d_x}} \min_{y \in \mathbb{R}^{d_y}} f(x,y) + g_1(x) + g_2(y), \tag{4}$$

where g_1 and g_2 can also be interpreted as regularizers or indicators. If we put $F(z) := F(x,y) = [\nabla_x f(x,y), -\nabla_y f(x,y)]$ and $g(z) = g(x,y) = g_1(x) + g_2(y)$, then it can be proved that $z^* \in \text{dom}\, g$ is a solution for (1) if and only if $z^* \in \text{dom}\, g$ is a solution for (4).

While minimization problems are widely investigated separately from variational inequalities, saddles are very often studied together with variational inequalities. In particular, lower bounds for the former are also valid for the latter. Moreover, upper bounds for variational inequalities are valid for saddle point problems. However, what is perhaps more important is that these lower and upper bounds match. This is in contrast to minimization, where the lower bounds are weaker.

2.3 Assumptions

Assumption 1 (Lipschitzness). *The operator F is L-Lipschitz continuous, i.e. for all $u, v \in \mathbb{R}^d$ we have*

$$\|F(u) - F(v)\| \leq L\|u - v\|. \tag{5}$$

For problems (3) and (4), L-Lipschitzness of the operator means that the functions $f(z)$ and $f(x,y)$ are L-smooth.

Assumption 2 (Strong monotonicity). *The operator F is μ-strongly monotone, i.e. for all $u, v \in \mathbb{R}^d$ we have*

$$\langle F(u) - F(v); u - v \rangle \geq \mu\|u - v\|^2. \tag{6}$$

For problems (3) and (4), strong monotonicity of F means strong convexity of $f(z)$ and strong convexity-strong concavity of $f(x,y)$.

Assumption 3 (δ-relatedness). *Each operator F_m is δ-related. It means that each operator $F_m - F$ is δ-Lipschitz continuous, i.e. for all $u, v \in \mathbb{R}^d$ we have*

$$\|F_m(u) - F(u) - F_m(v) + F(v)\| \leq \delta\|u - v\|. \tag{7}$$

While Assumptions 1 and 2 are basic and widely known, Assumption 3 requires further comments. This assumption goes back to the conditions of data similarity.

In more detail, we consider distributed minimization (3) and saddle point (4) problems:

$$f(z) = \frac{1}{M} \sum_{m=1}^{M} f_m(z), \quad f(x,y) = \frac{1}{M} \sum_{m=1}^{M} f_m(x,y),$$

and assume that for minimization local and global hessians are δ-*similar* [25, 32, 45, 52, 53]:

$$\|\nabla^2 f(z) - \nabla^2 f_m(z)\| \le \delta,$$

or for saddles second derivatives are differ by δ [13, 32]:

$$\|\nabla_{xx}^2 f(x,y) - \nabla_{xx}^2 f_m(x,y)\| \le \delta,$$
$$\|\nabla_{xy}^2 f(x,y) - \nabla_{xy}^2 f_m(x,y)\| \le \delta,$$
$$\|\nabla_{yy}^2 f(x,y) - \nabla_{yy}^2 f_m(x,y)\| \le \delta.$$

It turns out that if we look at machine learning and the data is u distributed between devices, it can be proven [25, 50] that $\delta = \tilde{\mathcal{O}}\left(\frac{L}{\sqrt{b}}\right)$, where b is the number of local data points on each of the workers.

3 Main Part

Our new algorithm Optimistic MASHA, as well as MASHA [11](the only compressed algorithm for variational inequalities already presented in the community), is based on the ideas of negative momentum and variance reduction technique [2, 33].

At the beginning of each iteration of Optimistic MASHA, each device sends the compressed version of δ_m^k to the server, and the server does a reverse broadcast. It is also possible to compress the messages coming from the server to the devices, but in practical cases there is very often no need for compression and the transfer process from the server takes less time than the sending from the devices to the server [3, 10, 37]. Also, the workes receive a bit of information b_k. This bit is generated randomly on the server and is equal to 1 with probability γ (where γ is small). Note that b_k can be generated locally, it is enough to use the same random generator and set the same seed on all devices. Then, all devices make a final update on z^{k+1} using Δ^k. One can notice that to compute Δ^k we need to know F, the full operator over all nodes. Then at first glance it seems that we need to always send the uncompressed operators. But that is not the case. It is enough to look at the w^{k+1} update. We put $w^{k+1} = z^{k+1}$ if the $b_k = 1$ or save it from the previous iteration $w^{k+1} = w^k$. In the case when the point $w^{k+1} = z^{k+1}$, we need to exchange the full values of $F_m(w^{k+1})$ in order that at the next iteration the value of $F(w^{k+1})$ is known to all nodes, but we do it rarely, with a small probability γ.

Unlike MASHA, we don not use arbitrary Q_m compressors on the devices, but a specific set $\{Q_m\}$, the so-called Permutation compressors, introduced in [47].

Definition 1 (Permutation compressors [47])

- **for** $d \geq M$. Assume that $d \geq M$ and $d = qM$, where $q \geq 1$ is an integer. Let $\pi = (\pi_1, \ldots, \pi_d)$ be a random permutation of $\{1, \ldots, d\}$. Then for all $u \in \mathbb{R}^d$ and each $m \in \{1, 2, \ldots, M\}$ we define

$$Q_m(u) := M \cdot \sum_{i=q(m-1)+1}^{qm} u_{\pi_i} e_{\pi_i}.$$

Algorithm 1. Optimistic MASHA

1: **Parameters:** Stepsize $\gamma > 0$, parameter τ, number of iterations K.
2: **Initialization:** Choose $z^0 = w^0 \in \mathcal{Z}$.
3: Server sends to devices $z^0 = w^0 = w^{-1}$ and devices compute $F_m(z^0)$ and send to server and get $F(z^0)$
4: **for** $k = 0, 1, 2, \ldots, K - 1$ **do**
5:    **for** each device m in parallel **do**
6:       Compute $F_m(z^k)$
7:       $\delta_m^k = F_m(z^k) - F_m(w^{k-1}) + \alpha[F_m(z^k) - F_m(z^{k-1})]$
8:       Send $Q_m\left(\delta_m^k\right)$ to server
9:    **end for**
10:   **for** server **do**
11:      Compute $\frac{1}{M} \sum_{m=1}^{M} Q_m(\delta_m^k)$ and send to devices
12:      Sends to devices b_k: 1 with probability γ, 0 with. probability $1 - \gamma$
13:   **end for**
14:   **for** each device m in parallel **do**
15:      $\Delta^k = \frac{1}{M} \sum_{m=1}^{M} Q_m^{\text{dev}}(\delta_m^k) + F(w^{k-1})$
16:      $z^{k+1} = \text{prox}_{\eta g}\left(z^k + \gamma(w^k - z^k) - \eta \Delta^k\right)$
17:      **if** $b_k = 1$ **then**
18:         $w^{k+1} = z^k$
19:         Compute $F_m(w^{k+1})$ and send it to server
20:         Get $F(w^{k+1})$ as a response from server
21:      **else**
22:         $w^{k+1} = w^k$
23:      **end if**
24:   **end for**
25: **end for**

- **for** $d \leq M$. Assume that $M \geq d$, $M > 1$ and $M = qd$, where $q \geq 1$ is an integer. Define the multiset $S := \{1, \ldots, 1, 2, \ldots, 2, \ldots, d, \ldots, d\}$, where each number occurs precisely q times. Let $\pi = (\pi_1, \ldots, \pi_M)$ be a random permutation of S. Then for all $u \in \mathbb{R}^d$ and each $m \in \{1, 2, \ldots, M\}$ we define

$$Q_m(u) := d u_{\pi_m} e_{\pi_m}.$$

The essence of such compressors is that their behavior is related to each other. For example, in the case when $d \geq M$ and $d = Mq$, each device transmits only q components of the full gradient and, importantly, these components are unique for this device. To make such a connection between compressors, one can set the same random seeds on the devices to generate permutations. The use of the Permutation compressors allows us to simultaneously benefit from both the data similarity and the compression of the transmitted information. Briefly, the idea can be described as follows. Because we have δ-related (δ-similar) operators $\{F_m\}$, in a rough approximation we can assume that $F_m \approx F$ and then $\delta_m^k \approx \delta^k = F(z^k) - F(w^{k-1}) + \alpha[F(z^k) - F(z^{k-1})]$. When we get $\{\delta_m^k\}$ compressed with the Permutation compressors, it is close to the necessary uncompressed operator $\frac{1}{M} \sum_{m=1}^{M} Q_m^{\text{dev}}(\delta_m^k) \approx \delta^k$. But in the meantime, we transmitted M times less information.

The following statements give a formal convergence of Algorithm 1. To begin, we give the lemma about the compressor from [47].

Lemma 1 (see [47]). *The Permutation compressors from Definition 1 are unbiased and satisfy*

$$\mathbb{E}\left[\left\|\frac{1}{M}\sum_{m=1}^{M} Q_m(a_m) - \frac{1}{M}\sum_{m=1}^{M} a_m\right\|^2\right] \leq \frac{1}{M}\sum_{m=1}^{M}\left\|a_m - \frac{1}{M}\sum_{i=1}^{M} a_i\right\|^2 \quad (8)$$

for all $a_1, \ldots, a_M \in \mathbb{R}^d$.

Next, we present the main theorem.

Theorem 1. *Consider the problem (1) under Assumptions 1, 2 and 3. Let $\{z^k\}$ be the sequence generated by Algorithm 1 with compressors from Definition 1 and parameters*

$$0 < \gamma \leq \frac{1}{8}, \quad \alpha = \frac{1}{2}, \quad \eta = \min\left\{\frac{\sqrt{\alpha\gamma}}{2\delta}, \frac{1}{8(L+\delta)}\right\}.$$

Then, given $\varepsilon > 0$, the number of iterations for $\|z^k - z^*\|^2 \leq \varepsilon$ is

$$O\left(\left[\frac{1}{\gamma} + \frac{L}{\mu} + \frac{\delta}{\sqrt{\gamma}\mu}\right]\log\frac{1}{\varepsilon}\right).$$

The proof of this Theoerm is given in the full version of the paper [9].

The resulting convergence estimate depends on the parameter γ. Let us find the way to choose it. In average, once per $\frac{1}{\gamma}$ iterations (when $b_k = 1$), we send uncompressed information. Hence, we can find the best option for γ. At each iteration the device sends $\mathcal{O}\left(\frac{1}{M} + \gamma\right)$ bits – each time information compressed by $\frac{1}{M}$ times and with probability γ we send the full package. We get the optimal choice of γ:

Corollary 1. *Under the conditions of Theorem 1, and with* $\gamma = \frac{1}{M}$, `Optimistic` `MASHA` *with the Permutation compressors has the following estimate on the total number of transmitted bits to find* ε-*solution*

$$O\left(\left[\frac{L}{M\mu} + \frac{\delta}{\sqrt{M}\mu}\right]\log\frac{1}{\varepsilon}\right).$$

Discussion of the Results in Terms of Compression. As noted earlier, under conditions of uniformly distributed data, the parameter $\delta = \tilde{O}\left(\frac{L}{\sqrt{b}}\right)$, where b is the number of local data points on each of the devices. Note that a typical situation is when $b \geq M$. Then the estimate from Corollary 1 can be rewritten as

$$O\left(\frac{L}{M\mu}\log\frac{1}{\varepsilon}\right).$$

State of the art methods for solving variational inequalities [12,28], which are also optimal algorithms in terms of the number of communications (but not the number of bits) give the next estimate on the number of bits

$$O\left(\frac{L}{\mu}\log\frac{1}{\varepsilon}\right).$$

`MASHA` can guarantee the following bound for the number of bits:

$$O\left(\frac{L}{\sqrt{M}\mu}\log\frac{1}{\varepsilon}\right).$$

This shows that our result is M times better than the uncompressed methods, and better than `MASHA` (which does not use δ-relatedness) by $\sqrt{M}$ times.

Discussion of the Results in Terms of Data Similarity. The algorithm from [13] using similarity for variational inequalities (in fact, for saddle point problems) has the following estimate for the number of bits to be forwarded

$$O\left(\frac{\delta}{\mu}\log\frac{1}{\varepsilon}\right).$$

If $M \geq b$, the estimate from the Corollary is transformed as follows

$$O\left(\left[\frac{L}{\sqrt{M}\sqrt{b}\mu} + \frac{\delta}{\sqrt{M}\mu}\right]\log\frac{1}{\varepsilon}\right) = O\left(\frac{\delta}{\sqrt{M}\mu}\log\frac{1}{\varepsilon}\right).$$

This result is $\sqrt{M}$ times better than from [13].

4   Experiments

The aim of our experiments is to test Corollary 1's result, namely the dependence of convergence on the parameter δ. To be able to vary the parameter δ we conduct our experiments on a distributed bilinear problem, i.e. the problem (4) with

$$f_m(x,y) := x^\top A_m y + a_m^\top x + b_m^\top y + \frac{\lambda}{2}\|x\|^2 - \frac{\lambda}{2}\|y\|^2, \tag{9}$$

where $A_m \in \mathbb{R}^{d \times d}$, $a_m, b_m \in \mathbb{R}^d$. This problem is λ-strongly convex–strongly-concave and, moreover, L-smooth with $L = \|A\|_2$ for $A = \frac{1}{M}\sum_{m=1}^{M} A_m$. We take $M = 10$, $d = 100$ and generate matrix A (with $\|A\|_2 \approx 100$) and vectors a_m, b_m randomly. Then we generate matrices B_m such that all elements of these matrices are independent and have an unbiased normal distribution with variance σ^2. We compute $A_m = A + B_m$. It may be considered that $\delta \sim \sigma$. In particular, we run three experiment setups: with small $\sigma \approx \frac{\|A\|_2}{100}$, medium $\sigma \approx \frac{\|A\|_2}{10}$ and big $\sigma \approx \|A\|_2$. λ is chosen as $\frac{\|A\|_2}{10^5}$.

We use the new Optimistic MASHA, the existing compression algorithm MASHA [11], and the classic uncompressed Extra Gradient [12,28] as competitors. In Optimistic MASHA and MASHA we use Permutation compressors. All methods are tuned as outlined in the theory of the corresponding papers.

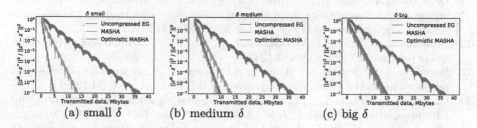

Fig. 1. Bilinear problem (9): Comparison of state-of-the-art methods with compression for variational inequalities for small, medium and big similarity parameters.

See Fig. 1 for the results. For small δ Optimistic MASHA is about $\sqrt{M}$ times superior to MASHA, and also outperforms the uncompressed method by a factor of M. With increasing δ Optimistic MASHA comes close to MASHA in its convergence.

5   Conclusion

In this paper, we considered distributed methods for solving the variational inequality problem. We presented a new method Optimistic MASHA. By combining two techniques: compression and data similarity, our method allows us to significantly reduce the number of information transmitted during communications. Experiments confirm the theoretical conclusions.

References

1. Agafonov, A., et al.: An accelerated second-order method for distributed stochastic optimization. arXiv preprint arXiv:2103.14392 (2021)
2. Alacaoglu, A., Malitsky, Y.: Stochastic variance reduction for variational inequality methods. arXiv preprint arXiv:2102.08352 (2021)
3. Alistarh, D., Grubic, D., Li, J., Tomioka, R., Vojnovic, M.: QSGD: communication-efficient SGD via gradient quantization and encoding. Adv. Neural Inf. Process. Syst., 1709–1720 (2017)
4. Arjevani, Y., Shamir, O.: Communication complexity of distributed convex learning and optimization. Adv. Neural Inf. Process. Syst. **28** (2015)
5. Bach, F., Jenatton, R., Mairal, J., Obozinski, G.: Optimization with sparsity-inducing penalties. arXiv preprint arXiv:1108.0775 (2011)
6. Bach, F., Mairal, J., Ponce, J.: Convex sparse matrix factorizations. arXiv preprint arXiv:0812.1869 (2008)
7. Bauschke, H., Combettes, P.: Convex Analysis and Monotone Operator Theory in Hilbert Spaces (2017). https://doi.org/10.1007/978-3-319-48311-5
8. Ben-Tal, A., Ghaoui, L.E., Nemirovski, A.: Robust Optimization. Princeton University Press, Princeton (2009)
9. Beznosikov, A., Gasnikov, A.: Compression and data similarity: Combination of two techniques for communication-efficient solving of distributed variational inequalities. arXiv preprint arXiv:2206.09446 (2022)
10. Beznosikov, A., Horváth, S., Richtárik, P., Safaryan, M.: On biased compression for distributed learning. arXiv preprint arXiv:2002.12410 (2020)
11. Beznosikov, A., Richtárik, P., Diskin, M., Ryabinin, M., Gasnikov, A.: Distributed methods with compressed communication for solving variational inequalities, with theoretical guarantees. arXiv preprint arXiv:2110.03313 (2021)
12. Beznosikov, A., Samokhin, V., Gasnikov, A.: Local sgd for saddle-point problems. arXiv preprint arXiv:2010.13112 (2020)
13. Beznosikov, A., Scutari, G., Rogozin, A., Gasnikov, A.: Distributed saddle-point problems under data similarity. Adv. Neural Inf. Process. Syst. **34** (2021)
14. Chambolle, A., Pock, T.: A first-order primal-dual algorithm for convex problems with applications to imaging. J. Math. Imaging Vision **40**(1), 120–145 (2011)
15. Chavdarova, T., Gidel, G., Fleuret, F., Lacoste-Julien, S.: Reducing noise in gan training with variance reduced extragradient. arXiv preprint arXiv:1904.08598 (2019)
16. Daneshmand, A., Scutari, G., Dvurechensky, P., Gasnikov, A.: Newton method over networks is fast up to the statistical precision. In: International Conference on Machine Learning, pp. 2398–2409. PMLR (2021)
17. Daskalakis, C., Ilyas, A., Syrgkanis, V., Zeng, H.: Training gans with optimism. arXiv preprint arXiv:1711.00141 (2017)
18. Esser, E., Zhang, X., Chan, T.F.: A general framework for a class of first order primal-dual algorithms for convex optimization in imaging science. SIAM J. Imaging Sci. **3**(4), 1015–1046 (2010)
19. Facchinei, F., Pang, J.: Finite-Dimensional Variational Inequalities and Complementarity Problems. Springer Series in Operations Research and Financial Engineering. Springer, New York (2007). https://books.google.ru/books?id=lX_7Rce3_Q0C
20. Facchinei, F., Pang, J.S.: Finite-Dimensional Variational Inequalities and Complementarity Problems. Springer Series in Operations Research. Springer, Heidelberg (2003). https://doi.org/10.1007/978-0-387-21815-1_6

21. Ghosh, A., Maity, R.K., Mazumdar, A., Ramchandran, K.: Communication efficient distributed approximate Newton method. In: IEEE International Symposium on Information Theory (ISIT) (2020). https://doi.org/10.1109/ISIT44484. 2020.9174216

22. Gidel, G., Berard, H., Vignoud, G., Vincent, P., Lacoste-Julien, S.: A variational inequality perspective on generative adversarial networks. arXiv preprint arXiv:1802.10551 (2018)

23. Goodfellow, I.J., et al.: Generative adversarial networks (2014)

24. Gorbunov, E., Burlachenko, K., Li, Z., Richtárik, P.: MARINA: faster nonconvex distributed learning with compression. In: 38th International Conference on Machine Learning (2021)

25. Hendrikx, H., Xiao, L., Bubeck, S., Bach, F., Massoulie, L.: Statistically preconditioned accelerated gradient method for distributed optimization. In: International Conference on Machine Learning. pp. 4203–4227. PMLR (2020)

26. Horváth, S., Kovalev, D., Mishchenko, K., Stich, S., Richtárik, P.: Stochastic distributed learning with gradient quantization and variance reduction. arXiv preprint arXiv:1904.05115 (2019)

27. Horváth, S., Richtárik, P.: A better alternative to error feedback for communication-efficient distributed learning. arXiv preprint arXiv:2006.11077 (2020)

28. Juditsky, A., Nemirovskii, A.S., Tauvel, C.: Solving variational inequalities with stochastic mirror-prox algorithm (2008)

29. Karimireddy, S.P., Rebjock, Q., Stich, S., Jaggi, M.: Error feedback fixes signsgd and other gradient compression schemes. In: International Conference on Machine Learning, pp. 3252–3261. PMLR (2019)

30. Konečný, J., McMahan, H.B., Yu, F., Richtárik, P., Suresh, A.T., Bacon, D.: Federated learning: strategies for improving communication efficiency. In: NIPS Private Multi-Party Machine Learning Workshop (2016)

31. Korpelevich, G.M.: The extragradient method for finding saddle points and other problems (1976)

32. Kovalev, D., Beznosikov, A., Borodich, E., Gasnikov, A., Scutari, G.: Optimal gradient sliding and its application to distributed optimization under similarity. arXiv preprint arXiv:2205.15136 (2022)

33. Kovalev, D., Beznosikov, A., Sadiev, A., Persiianov, M., Richtárik, P., Gasnikov, A.: Optimal algorithms for decentralized stochastic variational inequalities. arXiv preprint arXiv:2202.02771 (2022)

34. Li, Z., Kovalev, D., Qian, X., Richtárik, P.: Acceleration for compressed gradient descent in distributed and federated optimization. arXiv preprint arXiv:2002.11364 (2020)

35. Liang, T., Stokes, J.: Interaction matters: a note on non-asymptotic local convergence of generative adversarial networks. In: Chaudhuri, K., Sugiyama, M. (eds.) Proceedings of the Twenty-Second International Conference on Artificial Intelligence and Statistics. Proceedings of Machine Learning Research, vol. 89, pp. 907–915. PMLR (2019). https://proceedings.mlr.press/v89/liang19b.html

36. Mertikopoulos, P., Lecouat, B., Zenati, H., Foo, C.S., Chandrasekhar, V., Piliouras, G.: Optimistic mirror descent in saddle-point problems: Going the extra (gradient) mile. arXiv preprint arXiv:1807.02629 (2018)

37. Mishchenko, K., Gorbunov, E., Takáč, M., Richtárik, P.: Distributed learning with compressed gradient differences. arXiv preprint arXiv:1901.09269 (2019)

38. Nemirovski, A.: Prox-method with rate of convergence o (1/t) for variational inequalities with lipschitz continuous monotone operators and smooth convex-concave saddle point problems. SIAM J. Optim. **15**(1), 229–251 (2004)
39. Nemirovski, A.: Prox-method with rate of convergence o(1/t) for variational inequalities with lipschitz continuous monotone operators and smooth convex-concave saddle point problems. SIAM J. Optim. **15**, 229–251 (2004). https://doi.org/10.1137/S1052623403425629
40. Nesterov, Y.: Smooth minimization of non-smooth functions. Math. Program. **103**(1), 127–152 (2005)
41. Peng, W., Dai, Y.H., Zhang, H., Cheng, L.: Training gans with centripetal acceleration. Optim. Methods Softw. **35**(5), 955–973 (2020)
42. Philippenko, C., Dieuleveut, A.: Bidirectional compression in heterogeneous settings for distributed or federated learning with partial participation: tight convergence guarantees. arXiv preprint arXiv:2006.14591 (2020)
43. Reddi, S.J., Konečný, J., Richtárik, P., Póczós, B., Smola, A.: Aide: fast and communication efficient distributed optimization. arXiv preprint arXiv:1608.06879 (2016)
44. Richtárik, P., Sokolov, I., Fatkhullin, I.: EF21: a new, simpler, theoretically better, and practically faster error feedback. arXiv preprint arXiv:2106.05203 (2021)
45. Shamir, O., Srebro, N., Zhang, T.: Communication-efficient distributed optimization using an approximate newton-type method. In: Xing, E.P., Jebara, T. (eds.) Proceedings of the 31st International Conference on Machine Learning. Proceedings of Machine Learning Research, vol. 32, pp. 1000–1008. PMLR, Bejing (2014). https://proceedings.mlr.press/v32/shamir14.html
46. Smith, V., Forte, S., Ma, C., Takáč, M., Jordan, M.I., Jaggi, M.: CoCoA: a general framework for communication-efficient distributed optimization. J. Mach. Learn. Res. **18**, 1–49 (2018)
47. Szlendak, R., Tyurin, A., Richtárik, P.: Permutation compressors for provably faster distributed nonconvex optimization. arXiv preprint arXiv:2110.03300 (2021)
48. Tang, H., Yu, C., Lian, X., Zhang, T., Liu, J.: Doublesqueeze: parallel stochastic gradient descent with double-pass error-compensated compression. In: International Conference on Machine Learning, pp. 6155–6165. PMLR (2019)
49. Tian, Y., Scutari, G., Cao, T., Gasnikov, A.: Acceleration in distributed optimization under similarity. arXiv preprint arXiv:2110.12347 (2021)
50. Tropp, J.A.: An introduction to matrix concentration inequalities. arXiv preprint arXiv:1501.01571 (2015)
51. Verbraeken, J., Wolting, M., Katzy, J., Kloppenburg, J., Verbelen, T., Rellermeyer, J.S.: A survey on distributed machine learning. ACM Comput. Surv. **53**, 1–33 (2019)
52. Yuan, X.T., Li, P.: On convergence of distributed approximate newton methods: Globalization, sharper bounds and beyond. arXiv preprint arXiv:1908.02246 (2019)
53. Zhang, Y., Lin, X.: Disco: distributed optimization for self-concordant empirical loss. In: Bach, F., Blei, D. (eds.) Proceedings of the 32nd International Conference on Machine Learning. Proceedings of Machine Learning Research, vol. 37, pp. 362–370. PMLR, Lille (2015). https://proceedings.mlr.press/v37/zhangb15.html
54. Zheng, S., Huang, Z., Kwok, J.: Communication-efficient distributed blockwise momentum sgd with error-feedback. Adv. Neural Inf. Process. Syst. **32**, 11450–11460 (2019)

Game Theory and Mathematical Economics

Numerical Analysis of the Model of Optimal Savings and Borrowing

Alexey Chernov[1,2] and Aleksandra Zhukova[2]

[1] Moscow Institute of Physics and Technology, Dolgoprudny, Institute lane, 9,
141701 Moscow region, Russian Federation
[2] Federal Research Center "Computer Science and Control" of the Russian Academy
of Sciences, Vavilov Street, 44, Building 2, 119333 Moscow, Russian Federation
sasha.mymail@gmail.com

Abstract. This work presents the results of the numerical analysis of
the model of optimal savings and borrowing. The model is formulated
as the stochastic optimal control problem where the moments of time
when the state changes are random. The target (performance) functional
of the problem contains the penalty for the final values of the balance
account in the bank. This penalty has the form of the parameterized
softplus function. The sufficient optimality conditions contain the system
of functional and partial differential equations. The functional equation
is solved using the functional Newton method. The paper presents the
approach to numerical computation and the results of the numerical
solution to the optimal control problem.

Keywords: Optimal control · Functional equation · Newton's
method · Stochastic process

1 Introduction

Economic modelling uses models of optimal behavior of economic agents. Optimal control problems in economic models are often associated with description of uncertainty of the external environment or the internal decision-making process. This uncertainty is usually described by either a Wiener stochastic process or a Poisson process. Poisson uncertainty leads to optimality conditions, which often involve differential and difference equations that are difficult to investigate analytically using asymptotic methods or methods of perturbation theory. The possible method to solve them is numerical analysis, but the specifics of the equations being solved includes non-local expressions and special methods for solving them. This article presents a case of such non-local functional equation and an approach to obtain its solution.

The research of A. Chernov in Sects. 1.2 and 2.2 was supported by Russian Science
Foundation (project No. 21-71-30005). The research of A. Zhukova in Sects. 1.1 and
2.1 and 3 was supported by Russian Science Foundation (project No. 22-21-00746).

N. Olenev et al. (Eds.): OPTIMA 2022, LNCS 13781, pp. 165–176, 2022.
https://doi.org/10.1007/978-3-031-22543-7_12

1.1   Literature Review and Related Methods

A feature of optimal control (OC) problems in economic models is that they often make the assumption of concavity of the functional and the equations describing the dynamics. These assumptions usually guarantee the uniqueness of the solution or some convenient properties of the solution. With stochastic processes, analysis of optimal control problems is more complex. Optimality conditions contain second order partial derivatives in models with the Wiener process [10] and PDEs with differences (or shifts) for Poisson stochastic process [4,11]. The stochastic OC models in applied research are the building block of such widely used applied tool as the dynamic stochastic general equilibrium models. The main instrument to solve them is the dynamic programming method [8,12,13], but there are attempts to apply the Langrange multipliers' approach [2,14]. While in the deterministic case the optimality conditions would require to solve either the functional equation (dynamic programming), or the system of ODEs (Lagrange's method), in stochastic setting both methods give the expressions that define the control from functional equations of various complexity. The example of the functional equation is presented here and we propose an approach to solve it numerically.

1.2   Numerical Methods for Functional Equations

Numerical methods for functional equations, due to extreme complexity, require efficient numerical methods to analyze the solution. Our previous experience in applying numerical methods to computing the optimal control in the paper [15] highlights the usefulness of the functional Newton's method (2) [6], previously used for the one-dimensional case in the work [1].

The basic scheme of computation of the solution to the functional equation with the differential operator F

$$F(x) = 0 \tag{1}$$

is the Newton's method scheme

$$x^{k+1} = x^k - F'(x^k)^{-1}F(x^k). \tag{2}$$

It is important to note that the Newton's method requires to start the computation from good initial conditions and imposes strong requirements on the inverse operator $F'(\cdot)^{-1}$. One way to overcome this is to relax this requirement and, as a byproduct, to extend convergence area, by modifications:

1. Damped Newton's method;
2. Levenberg-Marquardt algorithm [9];
3. Quasi-Newton method [7].

The Damped Newton's Method. The first modification of the Newton's method can be written in the form (3)

$$x^{k+1} = x^k - \alpha_k F'(x^k)^{-1} F(x^k).$$ (3)

In the non-convergence area of the original Newton's method (2) the adaptive parameter α_k, known as a step size, can be selected to be less than 1 and may be calculated by the Armijo rule [5]. Such approach might allow one to achieve convergence of the method.

The Levenberg-Marquardt algorithm demonstrates advantage in the case of ill-posed or degenerate operator $F'(x^k)$, that might be important in case of the penalty method for the general problem. It can be written in the form, where the extra term $\alpha_k I$ is added to $F'(x^k)$

$$x^{k+1} = x^k - (\alpha_k I + F'(x^k))^{-1} F(x^k).$$ (4)

The parameter α_k might be selected according to some heuristics (or computation strategy) that enables convergence at each step of computation. In the case of $\alpha_k = 0$ the Newton's method step is done, while the relatively high value of the α_k changes the method to the gradient descent method.

The Quasi-Newton Method. The third option uses an approximation of the inverse operator $F'(x^k)^{-1}$ in the expression below

$$\begin{aligned} x^{k+1} &= x^k - H_k F(x^k), \\ H_0 &= I, \\ \lim_{k \to \infty} &\|H_k - F'(x^k)^{-1}\| = 0. \end{aligned}$$ (5)

The approximation of the inverse operator might start from some initial approximation (e.g. the identity operator in (5)) and is improved by the rule that selects the modification of the Quasi-Newton method (in classical optimization theory the Limited-memory Broyden-Fletcher-Goldfarb-Shanno algorithm [7]). The L-BFGS version is widely used nowadays.

Penalty approach is the another option (algorithm) that can be used to solve the problem considered in this article. In general, it can be written in the following form with the penalty model $M(x,t)$:

$$\begin{aligned} x^{k+1} &= \arg \min_x M(x, a_k); \\ a_{k+1} &= a_k \cdot m_k. \end{aligned}$$ (6)

In (6) the parameter m_k can be selected by different ways:

1. $m_k \equiv const$, where the value of the *const* can be 2, 3, ...;
2. $m_k = k$;
3. ...

One can combine iterative approaches using the Newton's method and the penalty approach as follows:

$$x^{k+1} = x^k - \alpha_k M'(x^k, a_k)^{-1} M(x^k, a_k);$$
$$a_{k+1} = a_k \cdot m_k. \tag{7}$$

This approach allows to improve convergence rate of the penalty method since it does not solve optimization problem on each step.

2   The Problem Statement

2.1   The Model of Optimal Control of Consumption and Borrowing

The model describes the optimal control of consumption and borrowing in a stochastic environment, where the optimal control enables changing the state variables only at random Poisson moments of time. The OC problem is the finite-horizon problem of optimal consumption $C(t) \geq 0$ and borrowing $K(t)$ by the economic agent (the consumer) that maximizes the functional

$$\mathrm{E}\left[\int_0^T U\left(\frac{C(t)}{C_0}\right) e^{-\Delta t} d\eta(t) - W\left(A(T) - L(T)\right)\right] \tag{8}$$

The outstanding debt $L(t)$ changes as the consumer takes a loan of the size $K(t)$.

$$dL(t) = K(t)\, d\eta(t), L(0) = L_0. \tag{9}$$

The balance of the bank account $A(t)$ changes as the consumer buys the consumption product in the amount $C(t)$ for the price $p_y(t)$, pays the interest at the rate $r_l(t)$ to the bank and receives the income as dividends from an enterprise $Z_\pi(t)$.

$$dA(t) = -p_y(t)\, C(t) d\eta(t) + K(t)\, d\eta(t)$$
$$- r_l(t)L(t)dt + Z_\pi(t)dt,\ A(0) = A_0. \tag{10}$$

The function $\eta(t)$ is the Poisson counting process $\eta(t, \omega)$ with ω omitted for convenience of notation. The intensity of the Poisson flow of events is constant, and denoted by Λ.

There are constraints on the state of the consumer: the debt is non-negative $L(t) \geq 0$, the solvency of the consumer is expressed as the inequality $A(t) \geq 0$, the ability to pay the debt should, ideally, hold by the end of the planning horizon $A(T) \geq L(T)$. Unfortunately, the latter is impossible to guarantee with certainty for the Poisson process. The controllability of the processes of $A(t)$ and $L(t)$ is imperfect, because by the final moment of time T with positive probability no transaction might take place. Instead, we introduce the penalty function $W(\cdot)$ in the form

$$W(x) = \frac{(\min(x, 0))^2}{2\, a^2},\ a \to 0. \tag{11}$$

The following assumptions are made in the model we consider in this work. The utility function $U(\cdot)$ is a monotone concave function with $U'(0) = \infty$. In particular, we assume $U(x) = ln(x)$. This form of utility function implies that the consumption rate $C(t)$ is strictly positive at any point of time. The functions $p_y(t), Z_\pi(t), r_l(t)$ are known and are assumed to be external to the consumer. They should be determined in the extended model of the economy, as the result of the general equilibrium, where households interact with firms, banks and the government. The non-anticipating stochastic processes $C(t), L(t), A(t), K(t)$ are assumed to be left-continuous with defined right limit and adpated to the natural filtration generated by the random Poisson process $\eta(t)$.

The analysis of the stochastic optimal control model involves formulation of the sufficient optimality conditions and equations to find the solutions in the form of synthesis (feedback). The solution to these conditions was found partially in [15], but not for the whole state space, due to challenges in the analysis of the functional equation. Recently we developed the approach to solving it and present the method in the sections below.

2.2   The Functional Equation for the Optimal Consumption

The main specifics of the optimal control problem on the finite planning horizon is the possibility of the boundary layer in the vicinity of the final moment of time. The model in [15] appears to have the boundary layer of the order $\frac{1}{A}$. Thus, in the time scale $t = T - \frac{\theta}{A}$ the functional equations that describe the consumption should satisfy the Eq. [15]

$$\min\left(0, \frac{-\sigma(\theta, Y) + Y}{a^2}\right) \tag{12}$$

$$= -\frac{e^{\theta - \Delta T}}{\sigma(\theta, Y)} + \int_0^\theta \frac{e^{\tau - \Delta T}}{\sigma(\tau, Y - \sigma(\theta, Y))} d\tau.$$

After rescaling this equation turns into

$$\frac{\min(0, x - f(t, x))}{a^2} \tag{13}$$

$$= -\frac{e^t}{f(t, x)} + \int_0^t \frac{e^\tau}{f(\tau, x - f(t, x))} d\tau.$$

The main computational challenge in solving this equations lays in the non-smooth left hand side of (13). To overcome this problem, we propose to use the smooth penalty function of the form called *softplus* [3] or *SmoothReLU*

$$-\frac{\ln(e^{kx} + 1)}{k} + x = -\frac{\ln(1 + e^{-kx})}{k}, \tag{14}$$

which we use either in the form of the left hand side of (14) or its right hand side, depending on the value of x, to avoid computing exponents of large numbers.

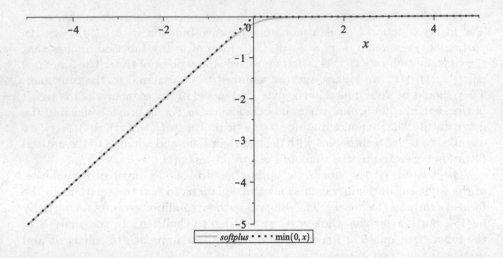

Fig. 1. The comparison between the softplus function for parameter $k = 4$ and the function $min(0, x)$.

The Fig. 1 demonstrates similarity of the two versions of the left hand side of the Eq. (12).

After substitution, we solve the equation

$$-\frac{\ln\left(e^{k\,(x - f(t,x))} + 1\right)}{k\,a^2} + \frac{x - f(t,x)}{a^2} \tag{15}$$

$$= -\frac{e^t}{f(t,x)} + \int_0^t \frac{e^\tau}{f(\tau, x - f(t,x))} d\tau.$$

An interesting question, important for the numerical analysis, is the convexity or concavity of the expressions in this equation. We left this question for the future research.

3   The Approach to Numerical Solution

We represent the Eq. (15) as the functional equation of the form

$$g(f) = 0 \tag{16}$$

and adapt the functional Newton method in the works [1, 15] to find its numerical solution.

The method implies updating the approximation according to the equation

$$f_n = f_{n-1} - \mathcal{F}^{-1}\left(f_{n-1}\right)\left[g\left(f_{n-1}\right)\right], \tag{17}$$

where we denote by $\mathcal{F}(f_{n-1})$ the inverse operator to the Frechet derivative of the operator $g(\cdot)$. In the Eq. (15), the Frechet derivative has the form

$$\mathcal{F}(f)[h]$$

$$= -\frac{\left(e^t + \int_0^t \frac{e^\tau D_2(f)(\tau, x - f(t,x))}{f(\tau, x - f(t,x))^2} d\tau f(t,x)^2\right)}{f(t,x)^2} h(t,x)$$

$$- \frac{h(t,x)}{a^2\left(e^{k\,x - k f(t,x)} + 1\right)} + \int_0^t \frac{e^\tau h(\tau, x - f(t,x))}{f(\tau, x - f(t,x))^2} d\tau. \qquad (18)$$

In order to be able to take the inverse of this operator, we take its approximation and omit the integral term

$$\int_0^t \frac{e^\tau h(\tau, x - f(t,x))}{f(\tau, x - f(t,x))^2} d\tau. \qquad (19)$$

This inevitably reduces the precision, but, as the experience with the similar equation, the work [15] shows that this modification preserves the convergence and the residual $g(f_n)$ becomes sufficiently small.

3.1 Asymptotic Solution for Large Negative x

We assume the asymptotic expansion of the function f as the series with respect to the variable x with coefficients as functions of t

$$f(t,x) = a_0(t) + \frac{a_1(t)}{x} + \frac{a_2(t)}{x^2} \qquad (20)$$

$$+ \frac{a_3(t)}{x^3} + \frac{a_4(t)}{x^4} + \frac{a_5(t)}{x^5} + O\left(t, x^{-6}\right). \qquad (21)$$

When substituted to (15) and after taking the series with respect to x of the right hand side and the left hand side, the coefficients of the exponents of x give integral equations on the functions $a_i(t)$. The even powers of x appear to correspond to zero coefficients $a_i(t), i = 0, 2, 4$.

$$\int_0^t \frac{e^\tau}{a_1(\tau)} d\tau = \frac{e^t}{a_1(t)} + a^{-2}, \qquad (22)$$

$$\int_0^t \frac{e^\tau a_3(\tau)}{a_1(\tau)^2} d\tau = -\frac{\left(-a_3(t) + a_1(t)^2\right) e^t}{a_1(t)^2}, \qquad (23)$$

$$\int_0^t \frac{e^\tau a_3(\tau)^2}{a_1(\tau)^3} d\tau = \int_0^t \frac{e^\tau a_5(\tau)}{a_1(\tau)^2} d\tau \qquad (24)$$

$$+ 2\frac{e^t a_3(t)}{a_1(t)} - a_1(t) e^t - \frac{e^t a_5(t)}{a_1(t)^2} + \frac{e^t a_3(t)^2}{a_1(t)^3},$$

$$\int_0^t \frac{e^\tau a_3(\tau)^3}{a_1(\tau)^4} d\tau = 2\int_0^t \frac{e^\tau a_3(\tau) a_5(\tau)}{a_1(\tau)^3} d\tau - 2 a_1(t)^2 e^t - 4\frac{e^t a_5(t)}{a_1(t)} \qquad (25)$$

$$+ 2\frac{e^t a_3(t)^2}{a_1(t)^2} + 6 e^t a_3(t) - 2\frac{e^t a_3(t) a_5(t)}{a_1(t)^3} + \frac{e^t a_3(t)^3}{a_1(t)^4}.$$

The solution to these equations gives the resulting asymptotic expression

$$f(t,x) = -\frac{a^2}{x} + \frac{a^4(t+1)}{x^3} - \frac{a^6(t+2)(2t+1)}{x^5}. \qquad (26)$$

This asymptotic expansion will serve as the boundary condition in the numerical analysis described below.

3.2   The Algorithm

As we describe above, we use the simplified operator instead of the Frechet derivative

$$\mathcal{F}_0(f)[h]$$

$$= -\left(\frac{e^t}{f(t,x)^2} + \int_0^t \frac{e^\tau f'_x(\tau, x - f(t,x))}{f(\tau, x - f(t,x))^2} d\tau \right) h(t,x)$$

$$- \frac{h(t,x)}{a^2 \left(e^{k\,x - k\,f(t,x)} + 1 \right)} \qquad (27)$$

and apply it to the iterative procedure we call the functional Newton's method

$$f_n = f_{n-1} - \alpha_n \mathcal{F}_0^{-1}(f_{n-1})\,[g(f_{n-1})]. \qquad (28)$$

One may see that the operator $\mathcal{F}_0(f)[h]$ is local with respect to (t,x). Therefore, its inverse is found by dividing 1 by the coefficient to h in (27).

$$\mathcal{F}_0^{-1}(f) \qquad (29)$$

$$= \left(-\frac{e^{4\,f(t,x) - 4\,x}}{a^2\left(1 + e^{4\,f(t,x) - 4\,x}\right)} - \frac{e^t}{f(t,x)^2} - \int_0^t \frac{e^\tau f'_x(\tau, x - f(t,x))}{f(\tau, x - f(t,x))^2} d\tau \right)^{-1}.$$

We propose the following approach to numerical solution of the Eq. (15). The iterative procedure (28) contains expressions with the target function with shifted argument $f(\tau, x - f(t, x))$, which makes the problem non-local. This means that, in order to find the $f(t, x)$ the procedure needs the data of the shifted value $f(\tau, x - f(t, x))$, $\tau \in [0, t]$ with the values $x - f(t, x)$ to the left of x (assuming the function f is non-negative, in accordance with the economic interpretation of the solution, as f defines the consumption).

Thus the value of the function f defined at the point $(\cdot, \tilde{x})$ depends on the values of the function f on the set $Q(\tilde{x}) = \{(t, x) \in R_+ \times R : x \le \tilde{x}\}$. Thus we can solve the problem in iterative manner in the domain of the function f. Let us consider the set of areas $Q_k = \{(t, x) \in R_+ \times R : x^k \le x \le x^{k+1}\}$, where $x^0 = -\infty$.

The boundary condition would provide the starting values of f at some range of x. In this role we propose the asymptotic approximation (26) for the large negative values of x.

The proposed approach starts with division of the range of values x into several intervals. The range of parameter t starts from 0 and ends with some value of order 10 to keep the t in the boundary layer.

Asymptotic Range of x, Range 0. The range $t \in [0, t_{max}]$, $x \in [-100, -80]$ with asymptotic values of $f(t, x)$ gives the matrix $f[1..i_{t_{max}}, 1..j_{max}, A]$ of values computed from the asymptotic formula (26).

Range 1. For the range $t \in [1, t_{max}]$, $x \in [-80, -20]$ the matrix $f[1..i_{t_{max}}, 1..j_{max}, 1]$ is computed iteratively using the procedure (28) and the values from the matrix with asymptotic range of x.

Range 2. For the range $t \in [0, t_{max}]$, $x \in [-20, 0]$ the matrix $f[1..i_{t_{max}}, 1..j_{max}, 2]$ is computed iteratively using the procedure (28) values from the matrices with asymptotic range of x and *Range 1*.

For the negative values of x we use the version of the softplus function (14)

$$-\frac{\ln\left(e^{k\,x} + 1\right)}{k} + x, \tag{30}$$

whereas for the positive range of x we use the version

$$-\frac{\ln\left(1 + e^{-k\,x}\right)}{k}. \tag{31}$$

Range 3. The range $t \in [0, t_{max}]$, $x \in [0, 10]$ seems to require more careful analysis. The partial solution to the initial Eq. (12) with certain rescaling in [15] demonstrates nonlinear growth of the function f in the range of small x. Therefore the grid might be made more finely-grained or even nonlinear as was realized in [15]. Here the the matrix $f[1..i_{t_{max}}, 1..j_{max}, 3]$ is computed using more fine grid on the dimension x using the iterative procedure (28) and values of f from the matrices of the asymptotic range of x, *Range 1* and *Range 2*.

Range 4. For the range $t \in [0, t_{max}]$, $x \in [10, 100]$ the solution was not computed in our previous attempt to approach numerically the Eq. (13) in [15], due to the nonlinearity of min in (13). In the present work the smooth approximation of the penalty function softplus enables using the procedure (28). The resulting matrix of the values f uses the data from the matrices *Range 0, 1, 2, 3.*

The last task is to combine the computed values of $f(t, x)$ in *Range 0-4* into one matrix with numerical solution of f.

3.3   Experiments

The simplest realization of the algorithm relies on the approximation of the integral by the midpoint rule and the partial derivative in (27) by the two-point formula. In fact, the derivative is taken for the values of $x - f(t, x)$ that might not correspond to any pint in the grid. For that, the nearest upper and lower values of x are found and the derivative is taken as the difference between f at the upper and lower value of x. For the integration, the values of $f(\tau, x - f(t, x))$ are also taken as the affine combination of the f at the nearest points on the grid, with weights defined by the distance to these points.

Experiments were conducted using the Maple computer algebra, to be able to conduct the first two iterations of the algorithm in the symbolic form. More iterations become impossible due to complexity of expressions in the Frechet derivative and the equation itself. Numerical computations show substantial values of the Frechet derivative (27) (of order 10^8) and high values of the residual (16) of order 10. As a result, the changes to the function f might be 10 or 100 times lager than its value (which, itself, is of order 10^{-3} to 10^{-4}. The way out is to, firstly, additionally multiply by f the $\mathcal{F}_0(f)[h]$ and $g(f)[h]$ as they computed separately. Secondly, the parameter α_k in the damped Newton's method is set to 10^{-3} for initial steps until the Frechet derivative and residual do not decrease substantially.

The penalty parameter a in the penalty function of the Eq. (12) we solve is set to 0.1 and the softplus parameter k is set to 4, but may take any other positive value.

It appears that for the range of $x = -20..0$ the convergence requires substantially adjusted parameter α_n as in the damped Newton's method (3). For the values $x < -20$ the parameter α_n can be set equalt to 1, but above at $x > -20$ the α_n that enables convergence keeping the values of $f(t, x) > 0$ is $\alpha_n = 0.01$ or 0.001. The resulting solution is presented in the Fig. 2.

What one may see is that the values of the solution are close to zero for negative x and as x grows the solution grows too. What is also important is that the solution f becomes close to x as t approaches zero. Notice that this time is scaled in the boundary layer so that the $t = 0$ means the end of the planning horizon. Economic interpretation is that the consumption spending is equal to all available money as the end of the planning horizon T comes. The consumption before the planning horizon might be seen from the graph for $t > 0$.

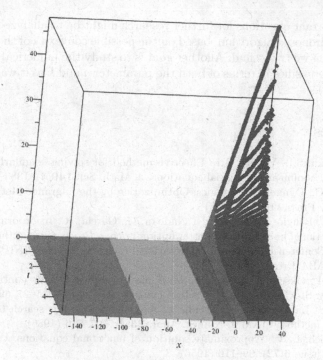

Fig. 2. The current result of computation in five regions, including asymptotic for large negative x. Each range is depicted in different color.

4   Discussion and Conclusions

This work presents the approach for numerical solution to the stochastic optimal control problem with state constraints in the boundary layer close to the end of the planning horizon. The results of the problem's analysis are listed below:

1. This work reveals the specifics of the functional equations that define the optimal control in the stochastic model with Poisson process, the non-local character of the equation.
2. The use of the approach based on the Newton's method to solve the functional equation became possible under the smoothness of the expressions in the equation and some boundary values of the solution that might be obtained by asymptotic expansion.
3. Due to the specific character of the equation, the derivative for the Newton method appears to be very complex and its inverse hard to find. The simplified version appears to deliver satisfactory results, perhaps at the cost of precision and low speed of convergence.
4. The results of the numerical experiments allow us to assume that the solution to the problem is convex, although the current quality of the result is far from excellent.

The important questions for further research might be to analyze the convergence of the proposed algorithm, based on the possible convexity of the functional or the function we try to find. Another goal is to study the numerical solution as the penalty intensifies in terms of both the parameter a and k as it was suggested in (6) and (7).

References

1. Bezrodnykh, S.I., Vlasov, V.I.: Effective method for solving singularly perturbed systems of nonlinear differential equations. J. Math. Sci. **149**(4), 1385–1399 (2008)
2. Chow, G.C.: Dynamic Economics: Optimization by the Lagrange Method. Oxford University Press, Oxford (1997)
3. Dugas, C., Bengio, Y., Bélisle, F., Nadeau, C., Garcia, R.: Incorporating second-order functional knowledge for better option pricing. Proceedings of the 13th International Conference on Neural Information Processing Systems (NIPS 2000), pp. 451–457. MIT Press (2000)
4. Elsanosi, I., Øksendal, B., Sulem, A.: Some solvable stochastic control problems with delay. Stochastics: Int. J. Prob. Stochastic Processes **71**(1–2), 69–89 (2000)
5. Grippo, L., Lampariello, F., Lucidi, S.: A nonmonotone line search technique for Newton's method. SIAM J. Numer. Anal. **23**(4), 707–716 (1986)
6. Kantorovich, L. V.: Approximate solution of functional equations. Uspekhi Mat. Nauk 11, Issue **6**(72), 99–116 (1956)
7. Liu, D.C., Nocedal, J.: On the limited memory BFGS method for large scale optimization. Math. Program. B. **45**, 503–528 (1989)
8. Maliar, L., Maliar, S., Winant, P.: Deep learning for solving dynamic economic models. J. Monetary Econ. **122**, 76–101 (2021)
9. Marquardt, D.W.: An algorithm for least-squares estimation of nonlinear parameters. J. Soc. Ind. Appl. Math. **11**(2), 431–441 (1963)
10. Øksendal, B.: Stochastic Differential Equations: An Introduction with Applications. Springer Science and Business Media, Heidelberg (2013). https://doi.org/10.1007/978-3-642-14394-6
11. Øksendal, B., Sulem, A.: Maximum principles for optimal control of forward-backward stochastic differential equations with jumps. SIAM J. Control Optim. **48**(5), 2945–2976 (2010)
12. Parra-Alvarez, J.C.: A comparison of numerical methods for the solution of continuous-time DSGE models. Macroecon. Dyn. **22**(6), 1555–1583 (2018)
13. Posch, O., Trimborn, T.: Numerical solution of dynamic equilibrium models under Poisson uncertainty. J. Econ. Dyn. Control **37**(12), 2602–2622 (2013). https://doi.org/10.1016/j.jedc.2013.07.001
14. Rong, S.: Theory of Stochastic Differential Equations with Jumps and Applications: Mathematical and Analytical Techniques with Applications to Engineering. Springer Science and Business Media, Heidelberg (2006). https://doi.org/10.1007/b106901
15. Zhukova, A., Pospelov, I.: Numerical analysis of the model of optimal consumption and borrowing with random time scale. In: Sergeyev, Y.D., Kvasov, D.E. (eds.) NUMTA 2019. LNCS, vol. 11974, pp. 255–267. Springer, Cham (2020). https://doi.org/10.1007/978-3-030-40616-5_19

Nash and Stackelberg Equilibria in Games with Pay-Off Functions Constructed by Minimum Convolutions of Antagonistic and Private Criteria

Victor Gorelik[1,2] and Tatiana Zolotova[3(✉)]

[1] FRC CSC RAS, Vavilova Street 40, 119333 Moscow, Russia
[2] Moscow Pedagogical State University, M. Pirogovskaya Street 1/1,
119991 Moscow, Russia
[3] Financial University under the Government of RF, Leningradsky Prospekt 49,
125993 Moscow, Russia
tgold11@mail.ru

Abstract. The paper proposes game models with pay-off functions being convolutions by the operation of taking minimum of two criteria one of which describes competition of players in some common (external) sphere of activity and the other describes private achievements of each player (in internal sphere). Strategies of players are distributions of resources between external and internal spheres. The first criterion of each player depends on strategies of all players; the second depends only on the strategy of given player. It is shown that under some natural assumptions of monotony of criteria such n-person games have good properties, namely, Nash equilibrium exists, is strong, stable and Pareto optimal. For two-person games, in Stackelberg equilibrium both the leader and the follower gain no less than in the best Nash equilibrium and the last belongs to γ-core.

Keywords: Pay-off functions · Stackelberg equilibrium · Nash equilibrium · Minimum convolutions · External sphere · Internal sphere

1 Introduction

Game theory is widely used to analyze political and military conflicts, economic competition, public goods and environmental negotiations, and so on. However, game-theoretic modeling faces a fundamental problem of the theory – the absence of a single principle of optimal behavior. This means that reasonable agreements from the point of view of one principle of optimality are not such from the point of view of another principle. As a result, an element of subjectivism is introduced into the study of real conflicts, associated with the choice of the concept of solution. One way out of this situation is to find such classes of game models for which different principles of optimality give the same solutions.

N. Olenev et al. (Eds.): OPTIMA 2022, LNCS 13781, pp. 177–191, 2022.
https://doi.org/10.1007/978-3-031-22543-7_13

As studies show, game models have good properties, characterized by the presence of equilibrium, stable and mutually beneficial compromises, if each participant has several criteria associated with different areas of activity. Apparently one of the first classes of such models was proposed by Germeier [1]. It served to describe conflicts, the participants of which, along with their, generally speaking, conflicting interests, have a common goal. Under certain conditions of monotony of objective functions and the use of the minimum type convolution, it was shown that in such games there is an equilibrium that is Pareto-optimal. Subsequently, these results were developed and generalized (for example, [2]).

At the same time, games with similar structures of interests arose when modeling the tasks of creating public goods and protecting the environment [3–23]. In particular, in the work of the authors of this paper, a game model with additive convolution of general and particular criteria was proposed, in which the Nash and Stackelberg equilibria exist and have good properties [24].

The question arises whether mutually beneficial and sustainable compromises are possible when there are no common interests, and moreover, there are antagonistic interests. At first glance, the answer must be negative, because in antagonistic games there is no room for compromise. However, this paper proposes a class of game models for which the answer to this question is positive, namely, the presence of additional private goals makes it possible to smooth out even a conflict situation with opposing interests. It seems that this result has an important practical interpretation, since allows us to hope for a peaceful resolution of military-political conflicts.

2   Formulation of the Model and Problem Statement

So, we consider a class of game models in which, on the one hand, it is possible to single out some common (external) sphere of activity in which the interests of all participants collide. In the case of two players, this sphere is described by an antagonistic game, and in the case of n persons, the objective function of each player depends on the strategies of all players, increases monotonically with an increase in his contribution to this sphere, and monotonically decreases with an increase in the contribution of each of the other participants. On the other hand, each participant has an internal sphere of activity in which one is completely independent from other players, i.e. the objective function for this sphere describes personal achievements, depends only on the strategy of the given player, and monotonically increases with increasing its contribution. The strategies of the players are the proportions of the distribution of resources between the external and internal spheres.

Denote the strategy of the i-th player (the share of funds invested in the inner sphere) by x_i, $0 \leq x_i \leq 1$, $x = (x_1, \ldots, x_n)$, $f_i(x_i)$ – the efficiency criterion of the i-th player, describing its achievements in the inner sphere, $g_i(x)$ is the efficiency criterion of the i-th player, describing its achievements in the external sphere. The payoff function of the i-th player, which one seeks to maximize, is

$$h_i(x) = \min\{f_i(x_i), g_i(x)\}. \tag{1}$$

Substantially, this means that the players measure their achievements in the internal and external spheres of activity and evaluate the situation according to the worst result. When defining a payoff function in the operation of taking the minimum in (1), one can introduce a weighting coefficient, however, due to the arbitrariness of the functions $f_i(x_i)$ and $g_i(x)$, its absence does not restrict the generality.

A model of this type can be interpreted in quite a variety of meaningful terms and refers to issues of economic rivalry in the world market, the struggle for political influence in various regions, the distribution of resources between the civilian and defense sectors of the economy, arms control negotiations, etc.

Concerning the functions $f_i(x_i)$ and $g_i(x)$, in addition to continuity, we accept the following assumptions, consistent with their meaning. Functions $f_i(x_i)$, generally speaking, it is natural to consider monotonically increasing, however, in order to be able to take into account the stimulating effect of the external sphere on the internal one, we will accept a somewhat more general assumption that $f_i(x_i)$ is a unimodal function with a maximum point x_i^+ (possibly $x_i^+{=}1$). The functions $g_i(x)$, generally speaking, must decrease in "their" variable x_i, since its increase means a decrease in the share of the contribution to the external sphere. However, if the resource is produced in the internal production sphere, then, in a certain range, the absolute value of the contribution to the external sphere can grow along with the growth of x_i. Therefore, we will consider $g_i(x)$ be unimodal as a function of x_i under fixed other arguments, with maximum points x_i^- (x_i^- does not depend on the value x_j, $j \neq i$, perhaps $x_i^-{=}0$). For the rest arguments $g_i(x)$ are assumed to be monotonically increasing, which corresponds to the opposite of interests (reducing the contribution of other players to the external sphere is beneficial for this player, since it increases its relative influence or competitiveness). Here we assume that $x_i^- < x_i^+$ and introduce the n-dimensional rectangular parallelepiped

$$P^{(n)} = \prod_{i=1}^{n} [x_i^-, x_i^+].$$

The game in normal form on the n-dimensional unit cube with payoff functions (1) under the assumptions made about the criteria $f_i(x_i)$ and $g_i(x)$ will be denoted by $H^{(n)}$.

Next, we consider two-person and n-person games with payoff functions (1) and investigate the existence and properties of Nash and Stackelberg equilibria.

3   Research of the Game $H^{(2)}$

Let us start with the question of the existence of Nash equilibrium situations.

Lemma 1. *In the game $H^{(2)}$ there is at least one situation of Nash equilibrium (x_1^N, x_2^N), all equilibria are strict, belong to $P^{(2)}$ and satisfy the following conditions:*

$$\left[f_i(x_i^N) - g_i(x_1^N, x_2^N) \right] \left(x_i^N - x_i^- \right) \left(x_i^N - x_i^+ \right) = 0, \ i = 1, 2.$$

Proof. The realization of the maximum of the function $h_i(x)$ with respect x_i for a fixed x_j $(i, j = 1, 2, \ i \neq j)$ is a single-valued continuous function $x_i^0(x_j)$, and due to the properties of the functions $f_i(x_i)$, $g_i(x)$ either $x_i^0(x_j) = x_i^-$, or $x_i^0(x_j) = x_i^+$, or $x_i^- < x_i^0(x_j) < x_i^+$ and $f_1(x_1^0(x_2)) = g_1(x_1^0(x_2), x_2)$ for the first player and $f_2(x_2^0(x_1)) = g_2(x_1, x_2^0(x_1))$ for the second player. The mapping $(x_1, x_2) \rightarrow \left(x_1^0(x_2), x_2^0(x_1)\right)$ of the unit square in the parallelepiped $P^{(2)}$ is single-valued and continuous, so by Brouwer's theorem it has fixed point, which is the Nash equilibrium. The remaining assertions of the lemma follow from the definition of equilibrium and the properties of the functions $x_i^0(x_j)$.

Remark 1. It follows from Lemma 1 that, without loss of generality, we can restrict ourselves to considering the game $H^{(2)}$ on $P^{(2)}$, where the functions $f_i(x_i)$ and $g_i(x)$ simply monotonically decrease or increase in the corresponding variables.

Consider now two games $\Gamma^{(1)}$ and $\Gamma^{(2)}$ with criteria $h_1(x_1, x_2)$, $h_2(x_1, x_2)$, in which, respectively, the first or the second player makes his move first (is the leader), and then, knowing this move, the other player (follower) makes his choice. Due to the uniqueness and continuity of the "response" function of the follower $x_i^0(x_j)$, such games have solutions (they are called Stackelberg equilibria).

The best result of the leader in the game $\Gamma^{(1)}$ (the first player) is

$$\gamma_1 = \max_{x_1^- \leq x_1 \leq x_1^+} h_1(x_1, x_2^0(x_1)),$$

and in the game $\Gamma^{(2)}$ (the second player) is

$$\gamma_2 = \max_{x_2^- \leq x_2 \leq x_2^+} h_2(x_1^0(x_2), x_2).$$

Let us denote the optimal leader strategies, realizing the maxima in these expressions, by x_1^{S0}, x_2^{S0}, and the answers of the followers by $x_2^{SV} = x_2^0\left(x_1^{S0}\right), x_1^{SV} = x_1^0\left(x_2^{S0}\right)$. The pair of strategies (x_1^{S0}, x_2^{SV}) is the Stackelberg equilibrium in the game $\Gamma^{(1)}$, and (x_1^{SV}, x_2^{S0}) is the Stackelberg equilibrium in the game $\Gamma^{(2)}$.

Lemma 2. *If in the game $H^{(2)}$ there are several situations of Nash equilibrium, then among them there is the best one for both players and in the games $\Gamma^{(1)}$ and $\Gamma^{(2)}$ both participants gain no less than in the best equilibrium situation.*

Proof. Let there be two Nash equilibrium situations (x_1^{N1}, x_2^{N1}) and (x_1^{N2}, x_2^{N2}). Functions $x_1^0(x_2), x_2^0(x_1)$ are non-decreasing on $P^{(2)}$.

Indeed, let $x_2^1 \geq x_2^2$. Then, due to the monotony properties of the function $g_1(x)$ on $P^{(2)}$ we have

$$g_1(x_1, x_2^2) \leq g_1(x_1, x_2^1) \ \forall x_1,$$

whence,

$$h_1(x_1, x_2^1) = \min\left\{f_1(x_1), g_1(x_1, x_2^1)\right\} \geq \min\left\{f_1(x_1), g_1(x_1, x_2^2)\right\} = h_1(x_1, x_2^2).$$

Let

$$\max_{x_1^- \leq x_1 \leq x_1^+,} h_1(x_1, x_2^1) = h_1(x_1^1, x_2^1),$$

$$\max_{x_1^- \leq x_1 \leq x_1^+,} h_1(x_1, x_2^2) = h_1(x_1^2, x_2^2),$$

i.e. $x_1^1 = x_1^0(x_2^1)$, $x_1^2 = x_1^0(x_2^2)$. Suppose that $x_1^1 < x_1^2$. Then we have the relations

$$h_1(x_1^1, x_2^2) \leq f_1(x_1^1) < f_1(x_1^2) = h_1(x_1^2, x_2^2),$$

and came up to a contradiction with the inequality

$$h_1(x_1, x_2^1) \geq h_1(x_1, x_2^2) \ \forall x_1,$$

because it follows

$$\max_{x_1^- \leq x_1 \leq x_1^+,} h_1(x_1, x_2^1) \geq \max_{x_1^- \leq x_1 \leq x_1^+,} h_1(x_1, x_2^2).$$

So the function $x_1^0(x_2)$ monotonically non decreasing in x_2. Similarly, it is proved that the function $x_2^0(x_1)$ monotonically non decreasing in x_1.

Therefore, taking into account the uniqueness of these functions, we have

$$x_1^{N1} < x_1^{N2} \ \leftrightarrow \ x_2^{N1} < x_2^{N2},$$

and due to the properties of monotony of functions $f_i(x_i)$, $g_i(x)$

$$h_i(x_1^{N1}, x_2^{N1}) < h_i(x_1^{N2}, x_2^{N2}), \ i = 1, 2.$$

The existence of the best Nash equilibrium (x_1^{N0}, x_2^{N0}) for both players with maximum coordinates follows from the continuity of the criteria.

In the games $\Gamma^{(1)}$ and $\Gamma^{(2)}$ the result of the leader is always no less than in any strict Nash equilibrium situation, and hence in the best one (x_1^{N0}, x_2^{N0}). For the follower this is not always the case, but for our games $x_i^{S0} \geq x_i^{N0}$, $i = 1, 2$. Indeed, assuming the contrary, i.e. $x_i^{S0} < x_i^{N0}$, we have $x_i^- < x_i^{N0}$, so $\gamma_i \geq h_i(x_1^{N0}, x_2^{N0}) = f_i(x_i^{N0}) > f_i(x_i^{S0}) \geq \gamma_i$ and arrive at a contradiction. Due to the fact that h_i does not decrease in x_j, $j \neq i$, the follower in the games $\Gamma^{(1)}$ and $\Gamma^{(2)}$ also gains at least as much as in the best Nash equilibrium situation.

Recall that the γ-core C_γ is the subset of the set of Pareto-optimal strategies for which

$$h_1(x_1, x_2) \geq \gamma_1, \quad h_2(x_1, x_2) \geq \gamma_2.$$

In our opinion, the following result is of greatest interest from the point of view of the properties of a two-person game.

Theorem 1. *If the function $g_0(x_1, x_2) = g_1(x_1, x_2) + g_2(x_1, x_2)$ does not increase by x_1, x_2 on $P^{(2)}$, in particular, $g_0(x_1, x_2) = const$, then in the game $H^{(2)}$ there is the unique Nash equilibrium, namely, (x_1^{N0}, x_2^{N0}), at that $(x_1^{N0}, x_2^{N0}) \in C_\gamma$ and is the only solution to the games $\Gamma^{(1)}$, $\Gamma^{(2)}$. If additionally $g_0(x_1, x_2)$ monotonically decreases in at least one argument, then (x_1^{N0}, x_2^{N0}) is the only element of γ-core.*

Proof. Assume the opposite, i.e. there are two equilibrium situations (x_1^{N1}, x_2^{N1}) and (x_1^{N2}, x_2^{N2}), at that $x_1^{N1} < x_1^{N2}$, $x_2^{N1} < x_2^{N2}$. Than $x_i^{N1} < x_i^+$ and

$$h_i(x_1^{N1}, x_2^{N1}) = g_i(x_1^{N1}, x_2^{N1}), \ i = 1, 2.$$

Therefore

$$\begin{aligned}
g_0(x_1^{N1}, x_2^{N1}) &= h_1(x_1^{N1}, x_2^{N1}) + h_2(x_1^{N1}, x_2^{N1}) \\
&< h_1(x_1^{N2}, x_2^{N2}) + h_2(x_1^{N2}, x_2^{N2}) \\
&\leq g_1(x_1^{N2}, x_2^{N2}) + g_2(x_1^{N2}, x_2^{N2}) = g_0(x_1^{N2}, x_2^{N2}),
\end{aligned}$$

and came to a contradiction. It means that equilibrium (x_1^{N0}, x_2^{N0}) is unique.

Let us show that it belongs to the set of Pareto-optimal strategies. Assume the opposite, i.e. for some pair of strategies $(x_1, x_2) \in P^{(2)}$ the following inequalities hold

$$h_1(x_1, x_2) \geq h_1(x_1^{N0}, x_2^{N0}), \quad h_2(x_1, x_2) \geq h_2(x_1^{N0}, x_2^{N0}),$$

and at least one of these inequalities is strict. If at the same time $x_i < x_i^{N0}$ for $i = 1$ or $i = 2$, than $x_i^- < x_i^{N0}$ and $h_i(x_1^{N0}, x_2^{N0}) = g_i(x_1^{N0}, x_2^{N0}) > h_i(x_1, x_2)$, came to a contradiction.

If $x_i > x_i^{N0}$, $i = 1, 2$, than $x_i^{N0} < x_i^+$ and

$$\begin{aligned}
h_1(x_1^{N0}, x_2^{N0}) + h_2(x_1^{N0}, x_2^{N0}) &= g_1(x_1^{N0}, x_2^{N0}) + g_2(x_1^{N0}, x_2^{N0}) \\
&= g_0(x_1^{N0}, x_2^{N0}) \geq g_0(x_1, x_2) \geq h_1(x_1, x_2) + h_2(x_1, x_2) \\
&> h_1(x_1^{N0}, x_2^{N0}) + h_2(x_1^{N0}, x_2^{N0}),
\end{aligned}$$

again came to a contradiction.

It remains to check two cases $x_1 > x_1^{N0}$, $x_2 = x_2^{N0}$ and $x_1 = x_1^{N0}$, $x_2 > x_2^{N0}$. In the first case we have $h_1(x_1, x_2) = h_1(x_1, x_2^{N0}) < h_1(x_1^{N0}, x_2^{N0})$ and came to a contradiction. Similarly, we came to a contradiction in the second case. It means, that equilibrium (x_1^{N0}, x_2^{N0}) belongs to the set of Pareto-optimal strategies.

Let us show, that (x_1^{N0}, x_2^{N0}) is the only solution to the game $\Gamma^{(1)}$.
Suppose, that $x_1^{N0} > x_1^{S0}$, than $x_1^{N0} > x_1^-$ and

$$\gamma_1 = h_1(x_1^{S0}, x_2^{SV}) \leq f_1(x_1^{S0}) < f_1(x_1^{N0})$$
$$= h_1(x_1^{N0}, x_2^{N0}) = h_1(x_1^{N0}, x_2^0(x_1^{N0})),$$

we came to a contradiction. Let now $x_1^{N0} < x_1^{S0}$, than $x_2^{SV} \geq x_2^{N0}$. If at the same time $x_2^{SV} = x_2^{N0}$, than $h_1(x_1^{N0}, x_2^{N0}) > h_1(x_1^{S0}, x_2^{N0}) = h_1(x_1^{S0}, x_2^{SV}) = \gamma_1$, we came to a contradiction.
  If $x_2^{SV} > x_2^{N0}$, than $x_2^{SV} > x_2^-$ and

$$h_2(x_1^{S0}, x_2^{SV}) = f_2(x_2^{SV}) > f_2(x_2^{N0}) \geq h_2(x_1^{N0}, x_2^{N0}),$$

whence due to Pareto-optimality (x_1^{N0}, x_2^{N0}) it follows

$$h_1(x_1^{S0}, x_2^{SV}) < h_1(x_1^{N0}, x_2^{N0}),$$

again came to a contradiction. So $x_1^{N0} = x_1^{S0}$, $x_2^{SV} = x_2^0(x_1^{N0}) = x_2^{N0}$. Similarly, it is proved that (x_1^{N0}, x_2^{N0}) is the only solution to the game $\Gamma^{(2)}$. Thus $\gamma_i = h_i(x_1^{N0}, x_2^{N0})$, $i = 1, 2$, i.e. $(x_1^{N0}, x_2^{N0}) \in C_\gamma$.
  It remains to show, that if the function $g_0(x_1, x_2\ )$ monotonically decreases in at least one argument, than (x_1^{N0}, x_2^{N0}) is the only element of γ-core. Assume the opposite, i.e. exists $(x_1, x_2) \in C_\gamma$, $(x_1, x_2) \neq (x_1^{N0}, x_2^{N0})$. Then, sequentially considering the cases $x_i < x_i^{N0}$ for $i = 1$ or $i = 2$, $x_i > x_i^{N0}$, $i = 1, 2$, $x_1 > x_1^{N0}$, $x_2 = x_2^{N0}$ and $x_1 = x_1^{N0}$, $x_2 > x_2^{N0}$, we arrive at a contradiction, as in the proof of Pareto-optimality (x_1^{N0}, x_2^{N0}).
  The theorem is completely proven.

Let us consider several numerical examples to illustrate the results and verify the significance of the conditions of Theorem 1.

Example 1. $f_i(x_i) = 2x_i$, $g_i(x) = 1 - x_i + x_j$, $i, j = 1, 2$, $i \neq j$.

Here the criteria satisfy the properties of the game $H^{(2)}$ and, moreover, are monotone, i.e. in this case $P^{(2)}$ coincides with the unit square. This means that there is at least one Nash equilibrium on it. It can be determined from the conditions $f_i(x_i) = g_i(x)$, $i = 1, 2$, and it is the situation $(0.5, 0.5)$. Moreover, in this example $g_0(x_1, x_2) = 2$, therefore, by virtue of Theorem 1, this equilibrium point is the unique one, Pareto-optimal and the only solution to the games Γ^1, Γ^2 (the leader and follower in these games have the optimal strategy $x_i^{S0} = 0.5$). However, $(0.5, 0.5)$ is not the only element in the set of Pareto-optimal strategies and γ-core. The set of Pareto-optimal strategies and γ-core contain any pair of strategies (x, x) under $x \geq 0.5$.
  Thus, the condition of monotonic decrease of the function $g_0(x_1, x_2)$ with respect to at least one argument in Theorem 1 is essential. Note that the game $H^{(2)}$ is essentially non-antagonistic despite the antagonism in the external sphere.

Example 2. $f_i(x_i) = 2x_i$, $g_i(x) = 1 - 2x_i + x_j$, $i, j = 1, 2$, $i \neq j$.

Here, too, $P^{(2)}$ coincides with the unit square, and the equilibrium is found from the conditions $f_i(x_i) = g_i(x)$, $i = 1, 2$, this is a pair of strategies $(1/3, 1/3)$. Moreover, in this example $g_0(x_1, x_2) = 2 - x_1 - x_2$, i.e. decreases monotonically in both arguments, so $(1/3, 1/3)$ is the only element of γ-core. However, it is not the only Pareto-optimal situation (for example, $(0.5, 1)$ is also a Pareto-optimal situation), but this was not stated in Theorem 1.

Example 3. $f_i(x_i) = 1 + x_i$, $g_i(x) = 1 - x_i + 2x_j$, $i, j = 1, 2$, $i \neq j$.

In this example $g_0(x_1, x_2) = 2 + x_1 + x_2$, i.e. increases in both arguments and does not satisfy the condition of Theorem 1. The Nash equilibrium is any point on the diagonal of the unit square, i.e., a pair of the form (x, x). The best equilibrium for both players is the situation $(1, 1)$, it is Pareto-optimal and the solution to the games Γ^1 and Γ^2. The rest of the equilibria are not Pareto-optimal.

Example 4. $f_i(x_i) = 3x_i$, $g_1(x) = 1 - x_1 + 2x_2$, $g_2(x) = 0.5 - 0.5x_2 + 2x_1$, $i = 1, 2$.

In this example $g_0(x_1, x_2) = 1.5 + x_1 + 1.5x_2$, i.e. increases in both arguments and does not satisfy the condition of Theorem 1. The Nash equilibrium here is the only one, it is $(0.45, 0.4)$, but it is not Pareto-optimal. Solutions to the games Γ^1 and Γ^2 are, respectively, $(1, 5/7)$ and $(3/4, 1)$, these situations are also not Pareto-optimal. The order of moves here is significant, namely, it is more profitable to be the follower than the leader, but both the leader and the follower win more than in the Nash equilibrium. However, the best for both players is the Pareto-optimal situation $(1, 1)$.

Example 5. $f_1(x_1) = ax_1$, $f_2(x_2) = kx_2$, $g_1(x) = a(1 - x_1) + x_2$, $g_2(x) = 1 - x_2 + x_1$.

Here, the parameter a can be interpreted as the amount of the resource at the first player (the second player has the amount of the resource equal to 1). We will assume that $a > 1$, i.e. the first player surpasses the second one in terms of available resources, and a is the measure of his superiority. The parameter k can be interpreted as a weighting factor that measures the achievements of the second player in the external and internal spheres, the larger k, the less importance he attaches to the internal sphere, we assume that $k > 1$. The function $g_0(x_1, x_2) = a + x_1 - ax_1$, it decreases by x_1 and does not depend of x_2.

This means that all the conditions of Theorem 1 are satisfied, the Nash equilibrium is the only one, Pareto-optimal, the only solution of the games Γ^1, Γ^2 and the only element of γ-core. Here also $P^{(2)}$ coincides with the unit square, and the equilibrium is found from the conditions $f_i(x_i) = g_i(x)$, $i = 1, 2$. Solving this system of equations, we obtain

$$x_1^{NO} = \frac{a(1+k)+1}{2a(1+k)-1}, \quad x_2^{NO} = \frac{3a}{2a(1+k)-1}.$$

At the equilibrium situation the payoffs in the external and internal spheres and the values of the general payoff functions of players are

$$h_1(x_1^{N0}, x_2^{N0}) = \frac{a[a(1+k)+1]}{2a(1+k)-1}, \quad h_2(x_1^{N0}, x_2^{N0}) = \frac{3ak}{2a(1+k)-1}.$$

Note, that $h_1(x_1^{N0}, x_2^{N0}) = h_2(x_1^{N0}, x_2^{N0})$ under $a = \frac{3k-1}{1+k}$.

Thus, in the competitive struggle in the external sphere, the player can, to a certain extent, compensate for the lack of resources by being less sensitive to the state of affairs in the internal sphere. For example, with $a = 2$ (twice superiority of the first player in terms of resources), equal results in the external sphere take place under $k = 3$.

The given examples show that the conditions imposed on the function $g_0(x_1, x_2)$ are essential, without them the assertions of Theorem 1 are false. These conditions have a clear substantive meaning: the individual achievements of participants in the external sphere decrease with an increase in the contribution of a competitor to it, but the total achievements with an increase in contributions can only grow (perhaps at the expense of other participants not explicitly described).

Next, we generalize some of the results to the n-person game.

4   Research of the Game $H^{(n)}$

Lemma 3. *In the game $H^{(n)}$ there is at least one Nash equilibrium x^N, all equilibria are strict, belong to the n-dimensional rectangular parallelepiped $P^{(n)}$ and satisfy the following conditions*

$$[f_i(x_i^N) - g_i(x^N)](x_i^N - x_i^-)(x_i^N - x_i^+) = 0, \ i = 1,\ldots,n.$$

Proof. The realization the maximum of the function $h_i(x)$ by x_i under fixed $x_j, j = 1,\ldots, n, j \neq i$, is a single-valued continuous function $x_i^0(x_{(-i)})$, $x_{(-i)} = (x_1,\ldots,x_{i-1}, x_{i+1},\ldots x_n)$, and due to the properties of the functions $f_i(x_i), g_i(x)$ true for anyone $x_{(-i)}$ either $x_i^0(x_{(-i)}) = x_i^-$, or $x_i^0(x_{(-i)}) = x_i^+$, or $x_i^- < x_i^0(x_{(-i)}) < x_i^+$ and $f_i(x_i^0(x_{(-i)})) = g_i(x_1,\ldots,x_{i-1}, x_i^0(x_{(-i)}), x_{i+1},\ldots x_n)$.

The map $(x_1,\ldots,x_n) \to (x_1,\ldots,x_{i-1}, x_i^0(x_{(-i)}), x_{i+1},\ldots x_n)$ of n- dimensional unit cube in $P^{(n)}$ is single-valued and continuous, so by Brouwer's theorem it has fixed point, which is the Nash equilibrium. The remaining assertions of the lemma follow from the definition of equilibrium and the properties of the functions $x_i^0(x_{(-i)})$.

Remark 2. It follows from Lemma 3 that, without loss of generality, we can restrict ourselves to considering the game $H^{(n)}$ on $P^{(n)}$, where the functions $f_i(x_i)$ and $g_i(x)$ simply monotonically decrease or increase by the corresponding variables.

Further we will need the following concept: a strong equilibrium in a game of n players is a situation, that for any coalition (a subset of the set of all players) there are no such strategies of its participants, the application of which, provided that the other players use equilibrium strategies, would give each participant in the coalition a payoff not less than the equilibrium value and at least one strictly greater. A strong equilibrium is both an ordinary Nash equilibrium (it corresponds to one-element coalitions in this definition) and a Pareto-optimal situation (it corresponds to the "grand" coalition of all players).

Theorem 2. *If the function*

$$g_0(x) = \sum_{i=1}^n g_i(x)$$

does not increase in all arguments, in particular, $g_0(x) = const$, then every equilibrium is strong. If the functions $g_i(x)$ can be represented in the form $g_i(x_i, \sum_{j \neq i} x_j)$, then there exists an equilibrium situation preferable for the "grand" coalition. If both conditions are fulfilled simultaneously, then the equilibrium situation is unique.

Proof. Let $x^N = (x_1^N, \ldots, x_n^N)$ be an arbitrary equilibrium situation and some coalition $K \subseteq \{1, 2, \ldots, n\}$ deviates from it, i.e. $x_i \neq x_i^N$, $i \in K$. If for at least one $i \in K$ fulfils $x_i < x_i^N$, then the payoff of the i-th player decreases in this case, since $x_i^- < x_i^N$ and for the new situation x the inequalities hold $h_i(x) \leq f_i(x_i) < f_i(x_i^N) = h_i(x_i^N)$, those coalition K as a whole does not benefit from such a deviation. Let now $x_i > x_i^N \ \forall \ i \in K$, then in the new situation, under the first condition of Theorem 2, we have $g_0(x) = \sum_{i=1}^n g_i(x) \leq g_0(x^N) = \sum_{i=1}^n g_i(x^N)$.

By virtue of the property of monotonous increase of the function $g_i(x)$ with respect to "foreign" arguments, the inequalities $g_i(x) > g_i(x^N) \ \forall \ i \in K$ are valid. So

$$\sum_{i \notin K} g_i(x) > \sum_{i \notin K} g_i(x^N),$$

which together with the previous inequality gives

$$\sum_{i \in K} g_i(x) < \sum_{i \in K} g_i(x^N).$$

So there is such a player number $i_0 \in K$, that $g_{i_0}(x) < g_{i_0}(x^N)$, and since $x_{i_0}^N < x_{i_0}^+$, than

$$h_{i_0}(x) \leq g_{i_0}(x) < g_{i_0}(x^N) = h_{i_0}(x^N),$$

those again, for the coalition K as a whole, such a deviation from equilibrium is not beneficial. The first assertion of the theorem is proved.

Let now the second condition of the theorem be satisfied and suppose that there are two different equilibrium situations $x^{N1} = (x_1^{N1}, \ldots, x_n^{N1})$ and

$x^{N2} = (x_1^{N2}, \ldots, x_n^{N2})$. Due to the monotony properties of the functions $f_i(x_i)$ and $g_i(x_i, \sum_{j\neq i} x_j)$ the realization the maximum of the function $h_i(x)$ by x_i is a single-valued continuous monotonically non-decreasing function of the form $x_i^0(\sum_{j\neq i} x_j)$. As $x_i^{N1} = x_i^0(\sum_{j\neq i} x_j^{N1})$, $x_i^{N2} = x_i^0(\sum_{j\neq i} x_j^{N2})$, then, taking into account the properties of the function $x_i^0(\sum_{j\neq i} x_j)$, we have the following implications.

If $\sum_{i=1}^n x_i^{N2} = \sum_{i=1}^n x_i^{N1}$, then $x_i^{N2} = x_i^{N1}$, $i = 1, \ldots n$, those equilibrium situations are the same. Indeed, suppose the opposite, namely, for some number i, the inequality $x_i^{N2} > x_i^{N1}$ is valid, than $\sum_{j\neq i} x_j^{N2} < \sum_{j\neq i} x_j^{N1}$ and we arrive at a contradiction with the monotone non-decreasing property of the function $x_i^0(\sum_{j\neq i} x_j)$.

If $\sum_{i=1}^n x_i^{N2} > \sum_{i=1}^n x_i^{N1}$, than $x_i^{N2} \geq x_i^{N1}$, $i = 1, \ldots n$, and exists i_0 such, that $x_{i_0}^{N2} > x_{i_0}^{N1}$.

Indeed, suppose the opposite, namely, for some number i, the inequality $x_i^{N2} < x_i^{N1}$ is valid, than $\sum_{j\neq i} x_j^{N2} > \sum_{j\neq i} x_j^{N1}$ and we arrive at a contradiction with the monotone non-decreasing property of the function $x_i^0(\sum_{j\neq i} x_j)$.

This means that if there are two different equilibrium situations, then all components of one of them are not less than the other and at least one component is strictly greater. At the same time, due to the monotonic properties of the criteria, for all players the first equilibrium situation is not worse than the second one, and at least for one player it is better. Among the equilibrium situations, there is one in which the sum of the components is strictly greater than in all the others. This equilibrium situation is preferable for the "grand" coalition (the only Pareto-optimal on the set of equilibrium situations).

If both conditions of the theorem are met, then on the one hand, each equilibrium is Pareto-optimal (as a strong equilibrium), and on the other hand, there can be only one Pareto-optimal among them, which means that the equilibrium is unique. The theorem has been proven.

Example 6. $f_i(x_i) = x_i$, $g_i(x) = (1 - x_i)[\sum_{j=1}^n (1 - x_j)]^{-1}, g_i(1, \ldots, 1) = 0$, $i = 1, \ldots, n$.

Both conditions of Theorem 2 are satisfied, which means that this game has a unique strict and strong equilibrium. Using the equilibrium conditions (Lemma 3), we obtain the system of equations

$$(1 - x_i)(n - \sum_{j=1}^n x_j)^{-1} = x_i, \ i = 1, \ldots, n.$$

The solution of this system gives the following equilibrium situation $x_i^N = \frac{1}{n}, i = 1, \ldots, n$.

We see that the more participants in this game, the more each player has to attract resources to the external sphere (you have to compete with many participants). Naturally, the possibility of cooperation is not assumed here.

5   Issues of Solution Stability According to Cournot

From a theoretical point of view, the considered class of game models has an ideal solution, since it meets all the basic principles of optimal behavior. However, in a practical conflict there may be various obstacles to reaching this theoretically justified compromise. First of all, they are related to the issues of awareness of the parties. Even the assumption of exact knowledge of one's capabilities and interests is not always justified, since each of the parties participating in the conflict can be a rather complex structure. However, we will adhere to the assumption accepted in game-theoretic models that each player knows exactly his objective function and the set of strategies. But the assumption that each player has complete information about the capabilities and especially the goals of other participants is, as a rule, unrealistic.

Consider some informational aspects for simplicity on the example of the game $H^{(2)}$. Knowing the qualitative properties of efficiency criteria (monotonity, unimodality) is a much weaker assumption than knowing the exact form of the other player's objective function. However, to establish an important property of the conflict described by the game $H^{(2)}$, namely, the existence of the best equilibrium situation for both participants, such information is sufficient. The information about the qualitative properties of the total payoff of the players in the external sphere, i.e. about monotony of the function $g_0(x_1, x_2)$, is also quite real. If it is known that this function does not increase in x_1, x_2, then we can conclude that there is a unique Nash equilibrium that is both a Stackelberg equilibrium and a Pareto-optimal situation (Theorem 1). However, to find this equilibrium, it is already necessary to know exactly the payoff functions of both players. There may be an exchange of information about these functions, but a possible bluff in order to achieve a better individual result cannot be ruled out.

So, let both players know all the specified qualitative properties of the criteria in the game $H^{(2)}$, but they know exactly only their payoff function. How can they come to an equilibrium situation? It turns out that the following iterative procedure of Cournot leads to equilibrium here, which can be interpreted as the result of successive individual actions or a negotiation process.

Let some situation (x_1^1, x_2^1) take place at the initial moment. The first player, knowing his payoff function $h_1(x_1, x_2)$ and, consequently, the best response function $x_1^0(x_2)$, unilaterally changes his strategy x_1^1 to $x_1^2 = x_1^0(x_2^1)$ (or proposes as a new compromise the situation (x_1^2, x_2^1)).

In response, the second player, knowing his payoff function $h_2(x_1, x_2)$ and, consequently, the best response function $x_2^0(x_1)$, changes his strategy x_2^1 to $x_2^2 = x_2^0(x_1^2)$. Then the process continues in the same way.

The result is a sequence

$$x_1^{k+1} = x_1^0(x_2^k), \quad x_2^{k+1} = x_2^0(x_1^{k+1}), \quad k = 1, 2, \ldots . \tag{2}$$

If this sequence converges to an equilibrium situation (x_1^N, x_2^N) for any initial situation (x_1^1, x_2^1), i.e. takes place

$$\lim_{k \to \infty} x_i^k = x_i^N, \ i = 1, 2,$$

then the equilibrium situation is called globally stable.

Theorem 3. *If the function* $g_0(x_1, x_2) = g_1(x_1, x_2) + g_2(x_1, x_2)$ *does not increase in* x_1, x_2 *on* $P^{(2)}$, *then the game* $H^{(2)}$ *has a unique Nash equilibrium* (x_1^N, x_2^N) *and it is globally stable.*

Proof. The existence of an equilibrium situation in the game $H^{(2)}$ and its uniqueness under the above assumption follows from Lemma 1 and Theorem 1. Let us show that it is globally stable.

Let two neighboring members of the sequence x_1^k satisfy the inequality $x_1^{k+1} \geq x_1^k$, then, due to the monotone nondecreasing function $x_2^0(x_1)$ (see the proof of Lemma 2), we have

$$x_2^{k+1} = x_2^0(x^{k+1}) \geq x_2^0(x_1^k) = x_2^k,$$

i.e. $x_2^{k+1} \geq x_2^k$.

Further, due to the monotone non-decreasing function $x_1^0(x_2^k)$, we have

$$x_1^{k+2} = x_1^0(x_2^{k+1}) \geq x_1^0(x_2^k) = x_1^{k+1},$$

i.e. $x_1^{k+2} \geq x_1^{k+1}$.

Hence, under the assumption made, the sequences x_i^k are monotonically nondecreasing, and since they are bounded (each belongs to the unit segment), there exist limits

$$\lim_{k \to \infty} x_i^k = x_i^L, \ i = 1, 2.$$

Passing to the limit in (2), taking into account the continuity of the functions $x_1^0(x_2)$, $x_2^0(x_1)$, we get

$$x_1^L = x_1^0(x_2^L), \quad x_2^L = x_2^0(x_1^L),$$

i.e. (x_1^L, x_2^L) is equilibrium. But according to the condition of the theorem, in the game $H^{(2)}$ there is a unique equilibrium situation (x_1^N, x_2^N), hence, $x_1^L = x_1^N$, $x_2^L = x_2^N$.

If we assume that $x_1^{k+1} \leq x_1^k$ is satisfied for two neighboring members of the sequence x_1^k, then the sequences $x_i^k, i = 1, 2$ will be monotonically non-increasing, and convergence to equilibrium is proved similarly.

Remark 3. Instead of (2), one can consider processes with a change of moves or simultaneous moves; they also converge to equilibrium.

Remark 4. Since under the conditions of Theorem 3 the Nash and Stackelberg equilibria coincide, the stability property extends to the latter.

6   Conclusion

Game models with pay-off functions being convolutions by the operation of taking minimum of two criteria, one of which describes competition of players in some common (external) sphere of activity and the other describes private achievements of each player (in internal sphere), are proposed. Summarizing the results, we note that for the chosen structure of the model and the monotony properties of the efficiency criteria, the solution of game problems simultaneously satisfies the basic principles of optimality, namely, the Nash and Stackelberg equilibria coincide and are Pareto-optimal situations. There is also no struggle for leadership, since equilibrium strategies are optimal for the leader and the follower in any order of moves. Moreover, due to the stability of the equilibrium, even individual actions of players who do not know or do not take into account the interests of other participants in the conflict lead, in the limit, to the same situations. These circumstances significantly increase the objectivity of resolving such conflicts on the basis of equilibrium and at the same time mutually beneficial compromises. The considered static models are supposed to be further developed for dynamic processes.

References

1. Germeier, Y.B., Vatel, I.A.: Games with hierarchical vector of interests. Isv. AN SSSR. Teknicheskaya kibernetika **1**(7), 54–69 (1974)
2. Kukushkin, N.S.: Strong Nash equilibrium in games with common and complementary local utilities. J. Math. Econ. **68**(1), 1–12 (2017)
3. Dixit, A.K., Nalebuff, B.J.: The Art of Strategy: A Game Theorist's Guide to Success in Business and Life. W.W. Norton Company, New York (2010)
4. Baliga, S., Maskin, E.: Mechanism design for the environment. In: Handbook of Environmental Economics, vol. 1, pp. 305–324. Elsevier, Amsterdam (2003)
5. Sefton, M., Shupp, R., Walker, J.M.: The effect of rewards and sanctions in provision of Public Good. Econ. Inq. **45**(4), 671–690 (2007)
6. Fehr, E., Gachter, S.: Cooperation and punishment in public goods experiments. Am. Econ. Rev. **90**(4), 980–994 (2000)
7. Hauert, C., Holmes, M., Doebeli, M.: Evolutionary games and population dynamics: maintenance of cooperation in public goods games. Proc. R. Soc. Lond. B Biol. Sci. **273**(1600), 2565–2571 (2006)
8. Zhang, J., Zhang, C., Cao, M.: How insurance affects altruistic provision in threshold public goods games. Sci. Rep. **5**, Article number: 9098 (2015)
9. Mu, Y., Guo, L.: Towards a theory of game-based non-equilibrium control systems. J. Syst. Sci. Complex. **1**(25), 209–226 (2012)
10. Mu, Y.F., Guo, L.: How cooperation arises from rational players? SCIENCE CHINA Inf. Sci. **56**(11), 1–9 (2013). https://doi.org/10.1007/s11432-013-4857-y
11. Mu, Y.: Inverse Stackelberg Public Goods Game with multiple hierarchies under global and local information structures. J. Optim. Theory Appl. **163**(1), 332–350 (2014)
12. Von, S.H.: Market Structure and Equilibrium. Springer, Heidelberg (2011). https://doi.org/10.1007/978-3-642-12586-7. Marktform und Gleichgewicht, Vienna, 1934. Translated by D. Bazin, L. Urch, R. Hill. (in English)

13. Basar T., Olsder, G.J.: Dynamic noncooperative game theory. In: The Society for Industrial Applied Mathematics. Academic Press, New York (1999)
14. Olsder, G.J.: Phenomena in inverse Stackleberg games, Part I: static problems. J. Optim. Theory Appl. **143**(3), 589–600 (2009)
15. Pang, J.S., Fukushima, M.: Quasi-variational inequalities, generalized Nash equilibria, and multi-leader-follower games. Comput. Manag. Sci. **2**(1), 21–56 (2005)
16. Luo, Z.Q., Pang, J.S., Ralph, D.: Mathematical Programs with Equilibrium Constraints. Cambridge University Press, Cambridge (1996)
17. Ye, J., Zhu, D.: New necessary optimality conditions for bilevel programs by combining the MPEC and value function approaches. SIAM J. Optim. **20**(4), 1885–1905 (2010)
18. Su, C.L.: Equilibrium problems with equilibrium constraints: stationarities, algorithms and applications. Ph.D. thesis, Stanford University, Stanford (2005)
19. Shen, H., Basar, T.: Incentive-based pricing for network games with complete and incomplete information. Adv. Dyn. Game Theory **9**, 431–458 (2007)
20. Staňková, K, Olsder, G.J., Bliemer, M.C.J.: Bi-level optimal toll design problem solved by the inverse Stackelberg games approach. WIT Trans. Built Environ. **89** (2006)
21. Staňková, K., Olsder, G.J., De Schutter, B.: On European electricity market liberalization: a game-theoretic approach. Inf. Syst. Oper. Res. **48**(4), 267–280 (2010)
22. Groot, N., Schutter, B.D., Hellendoorn, H.: On systematic computation of optimal nonlinear solutions for the reverse Stackelberg game. IEEE Trans. Syst. Man Cybern. Syst. **44**(10), 1315–1327 (2014)
23. Groot, N., Schutter, B.D., Hellendoorn, H.: Optimal affine leader functions in reverse Stackelberg games. J. Optim. Theory Appl. **168**(1), 348–374 (2016)
24. Gorelik, V., Zolotova, T.: Stackelberg and Nash equilibria in games with linear-quadratic payoff functions as models of public goods. In: Olenev, N.N., Evtushenko, Y.G., Jaćimović, M., Khachay, M., Malkova, V. (eds.) OPTIMA 2021. LNCS, vol. 13078, pp. 275–287. Springer, Cham (2021). https://doi.org/10.1007/978-3-030-91059-4_20

Fluctuations of Aggregated Production Capacity Near Balanced Growth Path

Nicholas Olenev[✉] [iD]

Federal Research Center "Computer Science and Control" of the
Russian Academy of Science, Moscow, Russia
nolenev@mail.ru

Abstract. The differential equation for the total production capacity in the vintage capacity model with age limit contains a delay. A characteristic solution to this equation is not only a balanced growth path, but also various kinds of oscillations near this path. The paper presents a multivalued solution for a special case of a fixed age limit and a given value of the share of new capacities. The state of such a system is determined by a whole segment of the trajectory.

Keywords: Aggregated production capacity · Vintage capacity model · Age limit · Balanced growth path · Fluctuations

1 Introduction

Descriptions of endogenous economic growth using a model with generations of physical capital (a vintage capital model) were proposed independently in the works of Leonid Kantorovich [9], Leif Johansen [8], and Robert Solow [18]. In the vintage capital models, technological progress concerns only the new generation of physical capital, thus, economic growth depends on the investment strategy.

Leif Johansen [7] introduced the concept of production capacity as the maximum possible output. This concept is clearer than the concept of physical capital. The concept of production capacity is more convenient to use in economic models to describe the production capabilities of production units, in particular, because it has the same dimension as output of production.

In the model with the age limit of capacities [12] dynamics of the total production capacity of economy is described by a functional differential equation with a positive delay. The golden rule of capacities accumulation for an endogenous production function is described in paper [13]. This golden rule gives a balanced growth path [17] of the model [12].

A number of monographs are devoted to the general theory of functional differential equations [2,5,10]. The book [16] describes the theory of boundary value problems for elliptic functional differential equations.

The publication has been prepared with the support of the "Research and development program using the infrastructure of the Shared Center of FRC CSC RAS".

N. Olenev et al. (Eds.): OPTIMA 2022, LNCS 13781, pp. 192–204, 2022.
https://doi.org/10.1007/978-3-031-22543-7_14

In numerical experiments with the model of the production function [12] on the data of Turkey [11], a characteristic solution of the model was discovered, which is an oscillation near the path of exponential growth.

Here we will study oscillations near a balanced growth path. To describe these oscillations, a complex form of representation of the growth rate of total production capacity is used. To solve the transcendental equation for the growth rate, the Lambert W function is used.

Section 2 describes dynamics of production capacities [12] and presents an optimal control problem to obtain a golden rule of growth [13]. Section 3 considers a differential equation for the total production capacity with a constant positive delay and introduces the procedure for finding its general solution. Section 4 presents the use of Lambert W function to write this general solution. Section 5 concludes the paper.

2   Problem Statement

Let t be the current time and τ be the time of creation of a production unit which has a production capacity $m(t, \tau)$. A micro description of the dynamics of production capacities with the limiting age $t - \tau \leq A(t)$ is given by Hypothesis 1 [12,13]. This hypothesis assumes that the number of jobs on a production unit is fixed from its creation up to reaching the age limit $A(t)$ while the capacity $m(t, \tau)$ decreases at a constant rate μ. If we additionally assume that at each time t all investments $J(t)$ are made in the best technology, then $m(t, t) = J(t)$ and

$$m(t, \tau) = J(\tau)e^{-\mu(t-\tau)}.$$

The best technology at time t is determined in the model by the least labor intensity $\nu(t)$, the rate of living labor costs per unit of output [12]. Then, by virtue of Hypothesis 1, the preservation of the number of jobs on the production unit created at time τ leads to the fact that its current labor intensity $\lambda(t, \tau)$ grows at the same rate μ, with which the production capacity $m(t, \tau)$ decreases,

$$\lambda(t, \tau) = \nu(\tau)e^{\mu(t-\tau)}$$

for $t - \tau \leq A(t)$.

The total capacity $M(t)$ of an industry or a national economy is determined by the capacity of production units with ages not exceeding the maximum permissible age $A(t)$.

$$M(t) = \int_{t-A(t)}^{t} m(t, \tau)d\tau.$$

Differentiating this expression with respect to time t according to the Leibniz integral rule [15], we obtain the following functional differential equation for the total production capacity $M(t)$ with the maximum capacity age $A(t)$ [12].

$$\frac{dM(t)}{dt} = J(t) - \mu M(t) - \left(1 - \frac{dA(t)}{dt}\right) J\left(t - A(t)\right) e^{-\mu A(t)}, \qquad (1)$$

where μ is the rate of degradation of production capacities due to wear and tear, and the value $J(t)$ is the speed of production capacity growth (the volume of investments in units of production capacity for a given period of time). The last term on the right side of the Eq. (1) describes the process of dismantling obsolete production capacities, that is, the production capacities whose age has exceeded the age limit $A(t)$. The following constraint on the age limit should be imposed in (1):

$$\frac{dA(t)}{dt} \leq 1,$$

which means that the once dismantled production capacities are not returned to the production process [14]. Scientific and technological progress can be described as a change of the least labor intensity

$$\frac{1}{\nu(t)} \frac{d\nu}{dt} = -\varepsilon \frac{J(t)}{M(t)},$$

where ε is the rate of scientific and technological progress, $0 < \varepsilon < 1$.

Note that the dynamics (1) of the total production capacity $M(t)$ taking into account the age limit $A(t)$ differs significantly from the dynamics of the total physical capital (capital stock) $K(t)$, usually used in standard macroeconomic models,

$$\frac{dK(t)}{dt} = I(t) - \kappa K(t),$$

where $I(t)$ is a new investment and κ is a depreciation rate of the total physical capital $K(t)$.

To write a golden rule of accumulation in terms of the total production capacity $M(t)$ instead of the total physical capital $K(t)$ the nation should maximize the intertemporal utility [13]

$$\int_0^\infty e^{-\delta t} u(c(t)) dt,$$

where δ is the discount rate and $u(c(t))$ is the instant utility of consumption.

In a closed economy the total output of production $Y(t)$ equals a sum of the final consumption expenditure $C(t)$ and the gross capital formation $I(t) = bJ(t)$, where b is the capital intensity factor of the new production capacities $J(t)$.

$$Y(t) = bJ(t) + C(t).$$

If $L(t)$ is the total labor (number of people employed in the economy), then $c(t) = C(t)/L(t)$ is the average per capita consumption,

$$x(t) = \frac{L(t)}{M(t)\nu(t)}$$

is the ratio of the average labor intensity $L(t)/M(t)$ to the least labor intensity $\nu(t)$, $f(x(t))$ is the homogeneous of degree one production function,

$$Y(t) = M(t)f(x(t)).$$

By definition of production capacity $M(t) \geq Y(t)$, so that the last function $f(x(t))$ has the meaning of the capacity utilization level, $f(x) \leq 1$.

Let's assume for simplicity that there is no scientific and technical progress, $\varepsilon = 0$, and the dynamics of the total production capacity (1) does not depend on the dismantling of obsolete capacities, for example, $dA/dt = 1$. Then the least labor intensity $\nu = const$, and

$$\frac{dM}{dt} = J(t) - \mu M(t),$$

so that the evolution of consumption depends on the next differential equation, the analogy of the key equation of the Solow model in terms of x

$$\dot{x}(t)/x(t) = \eta + \mu - (f(x(t)) - x(t)c(t)\nu)/b,$$

where the point above a variable is the derivative over time, and $\eta = \dot{L}/L$ is the growth rate of the total labor (or population).

The current value of the Hamiltonian

$$\Pi = u(c(t)) + \chi \left(\eta + \mu \right)x(t) - (x(t)f(x(t)) - x^2(t)c(t)\nu)/b \right),$$

where χ is an auxiliary variable. We can find

$$(u''(c)/u'(c))\dot{c} = \eta + \mu + \delta - (f'(x) - \nu c)x/b.$$

Indeed, $\partial H/\partial c(t) = u'(c) - \chi x^2\nu = 0$, $\dot{\chi} = \chi\delta - \partial H/\partial x = \chi(\delta - \eta - \mu + (f'(x) + xf'(x) - 2\nu xc)/b)$, so $\chi = u'(c)/(\nu x^2)$, and $\dot{\chi} = u''(c)\dot{c}/(\nu x^2) - 2\chi\dot{x}(t)/x(t)$. For stationary equilibrium we have $\dot{c} = 0$, $\dot{x} = 0$ and consequently

$$x^* f'(x^*) - f(x^*) = b\delta,$$
$$c^* = (f(x^*) - b(\eta + \mu))/(\nu x^*).$$

With this modified golden rule, the ratio of the average labor intensity to the least labor intensity, x, will be smaller because of the impatience of the society represented by the discount rate of time.

When studying the properties of the real economy, the Eq. (1) is interesting to study for the characteristic particular solutions of the economic model.

Functional differential equation (1) with a specified delay $A(t) > 0$, $dA/dt \leq 1$, can be solved (integrated) by the Bellman's method of steps [1,2,10], if the function $J(t)$ of the speed of increase in the production capacity $M(t)$ is specified, and, in addition, a prehistory of the change in the production capacity is specified, for example, on the interval $[-A, 0]$.

3   Delay Differential Equation for Production Capacity

Onward in the paper, we assume that the age limit A for production capacities is a positive constant. Then the Eq. (1) for the total production capacity $M(t)$ will be simplified to the following equation:

$$\frac{dM(t)}{dt} = J(t) - \mu M(t) - J(t - A)e^{-\mu A}. \tag{2}$$

For economic reasons, it should be assumed that $1 \ll A < \infty$.

In the works [12,13], some particular solutions of the delay differential equation (2) are studied, for which it is possible to obtain an analytical expression for the production function $f(t, x)$. In these solutions, in particular, the total production capacity $M(t)$ grows at a constant rate γ, which specified the balanced growth path of the model.

Here we consider an even more special case when not only the limiting age A, but also the ratio of new capacities $J(t)$ to total capacity $M(t)$ (the relative rate of increase in capacities) is constant:

$$\sigma(t) = J(t)/M(t) = \sigma = const.$$

The situation when investments are directly proportional to capacity is rare. However, even in this case, there are fluctuations around the balanced growth path. Let's look at this case here. The Eq. (2) gives a simple equation with a constant delay A for the total production capacity.

$$\frac{dM(t)}{dt} = (\sigma - \mu)M(t) - \sigma e^{-\mu A}M(t - A). \tag{3}$$

If the Eq. (3) holds for $t \geq t_0$ then for integrating it by the method of steps it is enough to know the history of the change in production capacities by time interval $[t_0 - A, t_0]$.

On the path of exponential growth the total production capacity $M(t)$ satisfies the next equality.

$$M(t) = M_0 e^{\gamma(t-t_0)}.$$

Then

$$M(t - A) = M_0 e^{\gamma(t-t_0)-\gamma A}, \frac{dM(t)}{dt} = \gamma M_0 e^{\gamma(t-t_0)},$$

and from here and (3) we have the next equation on γ

$$\gamma = \sigma - \mu - \sigma e^{-(\gamma+\mu)A}. \tag{4}$$

Denoting the dimensionless values

$$(\gamma + \mu)/\sigma = z, \sigma A = \alpha \tag{5}$$

we obtain from (4) the following equation.

$$1 - z = e^{-\alpha z}. \tag{6}$$

To study the stability of this particular golden rule solution [13], in this paper we consider complex value for the growth rate (4) of the total production capacity $M(t)$,

$$\gamma = \gamma_0 + i\gamma_1.$$

Any harmonic oscillation of type $x = a\cos(\omega t + \alpha)$ can be represented as

$$x = a\cos(\omega t + \alpha) = Re[ae^{i(\omega t + \alpha)}].$$

This representation, in particular, allows you to add harmonic oscillations.

Recording the growth rate in complex form will allow us to describe periodic oscillations in dynamics of the total production capacity. The value γ_0 is its real part of the constant real rate of economic growth, the value γ_1 is the real number, which is responsible for fluctuations near the exponential growth path.

The dimensionless value z is also complex,

$$z = z_0 + iz_1.$$

In accordance with the first equality in (5)

$$\gamma_0 = \sigma z_0 - \mu, \gamma_1 = \sigma z_1. \tag{7}$$

So we have proven the next

Theorem 1. *The components of the total production capacity $M(t)$ growth rate $\gamma = \gamma_0 + i\gamma_1$ are determined by the relations (7), $\gamma_0 = \sigma z_0 - \mu, \gamma_1 = \sigma z_1$. The complex number $z = z_0 + iz_1$ of the first equality in (5) is determined by the Eq. (6), $1 - z = e^{-\alpha z}$, where according to the last equality in (5) $\alpha = \sigma A > 0$.*

In real variables z_0, z_1, the relation (6) is written as

$$e^{-\alpha z_0}e^{-i\alpha z_1} = 1 - z_0 - iz_1.$$

According to Euler's formula for real value αz_1, $e^{-i\alpha z_1} = cos(\alpha z_1) + isin(\alpha z_1)$, we have the next system

$$e^{-\alpha z_0}cos(\alpha z_1) = 1 - z_0, \tag{8}$$

$$e^{-\alpha z_0}sin(\alpha z_1) = z_1. \tag{9}$$

Excluding z_0 from (9), we have

$$z_0 = \frac{1}{\alpha}ln\frac{sin(\alpha z_1)}{z_1}. \tag{10}$$

Substituting (10) into (8) we have the equation on z_1:

$$z_1 cot(\alpha z_1) = 1 - \frac{1}{\alpha}ln\frac{sin(\alpha z_1)}{z_1}. \tag{11}$$

Denoting $\alpha z_1 = y$ we get from (11)

$$\frac{sin(y)}{y} = \frac{e^\alpha}{\alpha}e^{-ycot(y)}. \tag{12}$$

From this Eq. (12) and Theorem 1 immediately follows the next theorem on the stability of balanced growth path for this model of production capacity.

Theorem 2. *Exponential growth of the model (3),*

$$dM(t)/dt = (\sigma - \mu)M(t) - \sigma e^{-\mu A} M(t - A),$$

for the total production capacity $M(t)$ is stable with real growth rate $\gamma = -\mu < 0$ if $\alpha = 1$ or, what the same, when $\sigma = 1/A$.

Proof. Indeed, in accordance with [2], the solution of the delay differential equation is stable if the imaginary part is equal to 0. Therefore, we need to find from (12) conditions when $y = 0$, because in this case by definition of y the value $z_1 = y/\alpha = 0$, and by condition (7) of Theorem 1 the value $\gamma_1 = \sigma z_1 = 0$.

We have

$$\lim_{y \to 0} \frac{sin(y)}{y} = 1.$$

It can be shown by L'Hopital's rule that

$$\lim_{y \to 0} e^{-ycot(y)} = \frac{1}{e}.$$

Indeed, in the exponential

$$-\lim_{y \to 0} \frac{ycos(y)}{sin(y)} = -\frac{cos(0) - sin(0)}{cos(0)} = -1.$$

According to (5) $\alpha = \sigma A$. It is a real positive value because the share of new capacities $\sigma > 0$, and the age limit $A \gg 1$. So we have from (12) that $\alpha = 1$. In this case $\sigma = 1/A$.

Substituting $z_1 = 0$ and $\alpha = 1$ to (8) we have

$$e^{-z_0} = 1 - z_0.$$

This transcendental equation has only one real root, $z_0 = 0$. According to (7) the root gives growth rate $\gamma = \gamma_0 = -\mu < 0$. Q.E.D.

It follows from the Theorem 2 that a stable, balanced growth path without fluctuations is an extremely rare phenomenon.

According to paper [12], the Russian economy 1970–2017 is characterized by the maximum age of production capacities $A = 25$ and the natural share of new capacities $\sigma = 0.11$. Then for this case $\alpha = \sigma A = 2.75$ and $a = e^{\alpha}/\alpha = 5.8823$. A solution of Eq. (12) for Russian data on the interval $y \in (2\pi, 3\pi)$ is $y^* = 7.37$ (see Fig. 1).

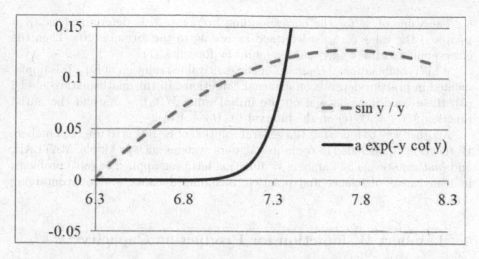

Fig. 1. A solution $y^* = 7.37$ of the Eq. (12) on the interval $y \in (2\pi, 3\pi)$ for the data of the Russian economy 1970–2017, $\sigma = 0.11, A = 25, \alpha = \sigma A, a = e^\alpha/\alpha$.

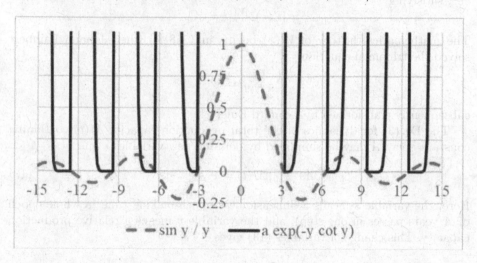

Fig. 2. All solutions of the Eq. (12) for the data of the Russian economy 1970–2017 with $a = e^\alpha/\alpha = 5.688$.

In the general case, a solution $y = \alpha z_1$ of Eq. (12) with $z_1 \neq 0$ can be find on the intervals $(\pi n, \pi(n+1)), n \in \mathbb{Z}$. The values $y = \pi n$ are poles for cotangent $cot(y)$.

The Eq. (12) has a multivalued solution. For $a = e$ there is a solution $y = 0$. For $0 < a < e$ there are two symmetric solutions on the interval $(-\pi, \pi)$. There are no solutions on the interval $(-\pi, \pi)$ for $a > e$ (see Fig. 2).

The value of z_1 for the corresponding branch of y is determined as $z_1 = y/\alpha$, and the value of z_0 is determed according to the formula (10). Then the corresponding values γ_0, γ_1 can be found by formulas (7).

Which combination of branches of the multivalued solution of Eq. (3) is implemented in practice depends on external conditions. In this mathematical model (3), these conditions depend on the initial value $M(t_0) = M_0$ and the initial function $M(t) = M_A(t)$ on the interval $t \in (t_0 - A, t_0)$.

Another way to represent the general solution of Eq. (3) is to use the Lambert W function [4], included in computer algebra systems such as Maple, MATLAB, and Mathematica. The Lambert W function has been applied to solve problems in many areas of science and practice, including dynamic systems containing delays.

4   Lambert W Function for Production Capacity

The Lambert W function, $W(z)$, is defined as the inverse of the function $g(z) = ze^z$, satisfying

$$W(z)e^{W(z)} = z.$$

The mathematical history of $W(z)$ begins in 1758 [4] when Johann Lambert solved the trinomial equation

$$z = q + z^m,$$

subsequently transformed by Leonard Euler.

The Eq. (3) for dynamics of the total production capacity $M(t)$ containing constant delay A may be simplified by defining new variables s and $v(s)$:

$$t = As + t_0, M(t) = v(s)e^{(\sigma-\mu)As}. \tag{13}$$

Here, the variable s means a dimensionless epoch-making time (a whole epoch of A years passes in one step), and the variable v means a relative production capacity. This change of variables (13) gives

$$\frac{dM(t)}{dt} = \frac{dv(s)}{ds}\frac{e^{(\sigma-\mu)As}}{A} + v(s)(\sigma - \mu)e^{(\sigma-\mu)As},$$

$$M(t - A) = v(s - 1)e^{(\sigma-\mu)A(s-1)}.$$

Substituting these relations into (3), we can consider the simple delay differential equation (DDE)

$$\frac{dv(s)}{ds} = -\beta v(s - 1) \quad (0 \le s < \infty), \tag{14}$$

given that $v(s) = \varphi(s)$ on $-1 \le s \le 0$ and a real positive dimensionless constant $\beta > 0$, $\beta = \sigma A e^{-\sigma A} = \alpha e^{-\alpha}$.

If the parameters σ, A have economic values then $0 < \sigma < 1$, and $A \gg 1$. If denote $w = -\sigma A$, then the inverse relation of the function

$$-\beta = g(w) = we^w$$

is the Lambert W function [3, 4, 19].

When share of new capacities $\sigma > 0$ and value $w = -\sigma A < 0$ are real numbers the equation $-\beta = we^w$ can be solved for real w only if $\beta < \frac{1}{e}$. In this case we get the two values $-\sigma A = w = W_0(-\beta)$ and $-\sigma A = w = W_{-1}(-\beta)$ if $0 < \beta < \frac{1}{e}$. Here W_0 is the principal branch of the Lambert W function.

If $\beta \to 0$ we get $-\sigma A = w = W_0(0) = 0$. If $\beta > \frac{1}{e}$ then w and β are complex numbers Then the equation $-\beta = we^w$ holds if and only if $w = W_k(-\beta)$ for some integer k.

To determine the solution (14) by the method of steps in the form of Myshkis [10], it is necessary to specify the historical function $\varphi(s)$ on the interval $-1 \leq s \leq 0$. In this case, there will be $v(0) = \varphi(0)$.

In particular, let the function be $\varphi(s) = re^{\rho s}$. If $\rho = 0$ ($\varphi(s) = r, -1 \leq s \leq 0$), for each s we have a solution similar to the one presented, for example, in [10]:

$$v(s) = r \sum_{j=0}^{n} \frac{(-\beta)^j (s - j + 1)^j}{j!} \quad (n - 1 \leq s \leq n; \ n = 1, 2, \ldots).$$

If $\rho \neq 0$ and $r > 0$ we successively get:

$0 \leq s \leq 1$:

$$\frac{dv(s)}{ds} = -\beta r e^{\rho(s-1)}; \quad v(s) = \left(-\frac{\beta}{\rho e^\rho}\right) r \left(e^{\rho s} - 1\right) + v(0);$$

$$v(0) = r; \quad v(s) = r \left(1 + \frac{\beta}{\rho e^\rho}\right) - \frac{\beta}{\rho e^\rho} r e^{\rho s}.$$

$1 \leq s \leq 2$:

$$\frac{dv(s)}{ds} = -\beta r \left(1 + \frac{\beta}{\rho e^\rho} - \frac{\beta}{\rho e^\rho} e^{\rho(s-1)}\right);$$

$$v(s) = -\beta r \left(1 + \frac{\beta}{\rho e^\rho}\right) (s - 1) + \left(-\frac{\beta}{\rho e^\rho}\right)^2 r \left(e^{\rho s} - e^\rho\right) + v(1);$$

$$v(1) = r \left(1 + \frac{\beta}{\rho e^\rho} - \frac{\beta}{\rho}\right);$$

$$v(s) = r \left(1 + \frac{\beta}{\rho e^\rho}\right) \left(1 + \beta - \frac{\beta}{\rho}\right) - \beta r \left(1 + \frac{\beta}{\rho e^\rho}\right) s + \left(\frac{-\beta}{\rho e^\rho}\right)^2 r e^{\rho s}.$$

$$\cdots\cdots\cdots$$

Using mathematical induction, a general formula can be obtained.

To determining the solution of DDE (14) immediately by use of the Lambert W function we can write [4]

$$v(s) = \sum_{n=-\infty}^{\infty} C_n e^{W_n(-\beta)s}, \qquad (15)$$

where values C_n are determined numerically by $\varphi(s)$ on $-1 \leq s \leq 0$.

If history of DDE (3) is determined by $M(t) = \psi(t)$ on $t \in [t_0 - A, t_0]$ then according to (13)

$$\varphi(s) = \psi(As + t_0)e^{-(\sigma-\mu)As}$$

on $-1 \leq s \leq 0$, and vice versa, if the historical function $\varphi(s)$ is given for DDE (14) then in the original variables the condition has the form $M(t) = \psi(t)$ on $t \in [t_0 - A, t_0]$, where

$$\psi(t) = \varphi\left(\frac{t - t_0}{A}\right)e^{(\sigma-\mu)(t-t_0)}.$$

Substituting (15) in (13) we have general solution for $M(t)$.

$$M(t) = \sum_{n=-\infty}^{\infty} C_n e^{W_n(-\beta)(t-t_0)/A} e^{(\sigma-\mu)(t-t_0)}. \tag{16}$$

Assuming that the initial function $\varphi(s)$ for DDE (14) is given on the interval $-1 \leq s \leq 0$ we can approximate the sum (15) by truncating it and matching the coefficients C_n with the function. So that, we look at the next sum

$$v(s) = \sum_{n=-(N+1)}^{N} C_n e^{W_n(-\beta)s} \tag{17}$$

and try to determine such C_n that make $v(s)$ as close to $\varphi(s)$ as we can on the initial interval.

We can determine the values of the C_n directly by using Least Squares trying to minimize the L_2-norm of the difference between $v(s)$ and $\varphi(s)$ on $-1 \leq s \leq 0$.

This problem for a given initial function can be solved numerically using the package of computer algebra (Maple, MATLAB or Mathematica). The solution of this problem in Maple is presented in [6] in the particular case of $\varphi = 1$ on the interval $-1 \leq s \leq 0$.

5   Conclusions

We know that an economic growth model is characterized by balanced growth path [17]. We study the characteristic mode of fluctuations near the exponential growth path of total production capacity $M(t)$ and investments $J(t)$ at a given rate:

$$M(t) = M_0 e^{\gamma(t-t_0)}, J(t) = J_0 e^{\gamma(t-t_0)}, \tag{18}$$

where $\gamma = \gamma_0 + i\gamma_1$.

A comprehensive representation of the growth rate of total capacity made it possible to describe fluctuations near balanced growth, which are often encountered in the economy.

A modern technique for finding a solution to a transcendental equation using the Lambert W function available in modern mathematical packages is presented.

In order to find the growth rate γ in expression (18), it is enough, in accordance with (13), to find a solution to Eq. (14) under the condition $v(s) = \varphi(s)$ on the historical interval $s \in [-1, 0]$.

A large variety of possible trajectories even in the case (18) is due to different initial functions.

References

1. Bellen, A., Zennaro, M.: Numerical Methods for Delay Differential Equations. Clarendon Press, Oxford (2003)
2. Bellman, R., Cooke, K.L.: Differential-Difference Equations. Academic Press, New York, London (1963)
3. Boivin, A., Zhong, H.: Completeness of systems of complex exponentials and the Lambert W functions. Trans. Am. Math. Soc. **359**, 1829–1849 (2007)
4. Corless, R.M., Gonnet, G.H., Hare, D.E.G., Jeffrey, D.J., Knuth, D.E.: On the Lambert W function. Adv. Comput. Math. **5**, 329–359 (1996)
5. Hale, J.: Theory of Functional Differential Equations. Springer, New York (1977). https://doi.org/10.1007/978-1-4612-9892-2
6. Heffernan, J.M., Corless, R.M.: Solving some delay differential equations with computer algebra. Math. Sci. **31**, 21–34 (2006)
7. Johansen, L.: Production functions and the concept of capacity. Collect. Econ. Math. Econom. **2**, 49–72 (1968)
8. Johansen, L.: Substitution versus fixed production coefficients in the theory of economic growth: a synthesis. Econometrica **27**(2), 157–176 (1959)
9. Kantorovich, L.V., Gorkov, L.I.: On some functional expressions that arise in the analysis of a single-product economic model. Doklady AN SSSR **129**(4), 732–735 (1959). (in Russian)
10. Kolmanovskii, V., Myshkis, A.: Applied Theory of Functional Differential Equations. Kluwer, Dordrecht (1992)
11. Olenev, N.: Identification of an aggregate production function for the economy of Turkey. In: New Trends in Economics and Administrative Sciences. Proceedings of the Izmir International Congress on Economics and Administrative Sciences (IZCEAS 2018), pp. 1761–1770. Detay Yayincilik, Izmir (2018)
12. Olenev, N.N.: Identification of a production function with age limit for production capacities. Math. Model. Comput. Simul. **12**(4), 482–491 (2020). https://doi.org/10.1134/S2070048220040134
13. Olenev, N.: Golden rule saving rate for an endogenous production function. In: Jaćimović, M., Khachay, M., Malkova, V., Posypkin, M. (eds.) OPTIMA 2019. CCIS, vol. 1145, pp. 267–279. Springer, Cham (2020). https://doi.org/10.1007/978-3-030-38603-0_20
14. Olenev, N.N., Petrov, A.A., Pospelov, I.G.: Model of change processes of production capacity and production function of industry. In: Samarsky, A.A., Moiseev, N.N., Petrov, A.A. (eds.) Mathematical Modelling: Processes in Complex Economic and Ecologic Systems, pp. 46–60. Nauka, Moscow (1986). https://doi.org/10.13140/RG.2.1.3938.8880. (in Russian)

15. Protter, M.H., Morrey, C.B.: Differentiation under the integral sign. Improper integrals. The Gamma function. In: Protter, M.H., Morrey, C.B. (eds.) Intermediate Calculus. Undergraduate Texts in Mathematics, pp. 421–426. Springer, New York (1985). https://doi.org/10.1007/978-1-4612-1086-3_8
16. Skubachevskii, A.L.: Elliptic Functional Differential Equations and Applications. Birkhäuser, Basel (1996)
17. Solow, R.M.: A contribution to the theory of economic growth. Q. J. Econ. **70**(1), 65–94 (1956)
18. Solow, R.M.: Investment and technical progress. In: Mathematical Methods in the Social Sciences, pp. 48–93. Stanford University Press, Stanford (1960)
19. Yi, S., Nelson, P.W., Ulsoy, A.G.: Time-Delay Systems: Analysis and Control Using the Lambert W Function. World Scientific, Singapore (2010)

Applications

On the Simultaneous Identification of the Volumetric Heat Capacity and the Thermal Conductivity of a Substance

Alla Albu[ID], Andrei Gorchakov[ID], and Vladimir Zubov[✉][ID]

Federal Research Center "Computer Science and Control" of the Russian Academy of Sciences, Moscow, Russia
vladimir.zubov@mail.ru

Abstract. The study of nonlinear problems associated with heat transfer in the substance is very important for practice. Previously, we proposed an efficient algorithm for identifying the thermal conductivity of a substance based on the results of experimental observation of the dynamics of the temperature field in an object. In this paper, we investigate the possibility of extending the application of the proposed algorithm to obtain a numerical solution to the problem of simultaneous identification of the temperature-dependent the thermal conductivity and the volumetric heat capacity of the substance under study. The consideration is carried out on the basis of the first initial-boundary value problem for the one-dimensional non-stationary heat equation. The considered inverse coefficient problem is reduced to a variational problem, which is solved by gradient methods based on the application of the Fast Automatic Differentiation technique.

Keywords: Nonlinear problems · Inverse coefficient problems · Heat equation · Numerical algorithm · Fast Automatic Differentiation

1 Introduction

In the description and mathematical modeling of many thermal processes, the classical heat equation is often used. The density of the substance, its specific thermal capacity, and the thermal conductivity appearing in this equation are assumed to be known functions of the coordinates and temperature. Additional boundary conditions make it possible to determine the dynamics of the temperature field in the substance under examination.

However, the substance parameters are not always known. It often happens that the volumetric heat capacity and the thermal conductivity depend only on

The work was carried out using the infrastructure of the Center for Collective Use "High-Performance Computing and Big Data" (CCU "Informatics") FRC CSC RAS (Moscow).

the temperature, and this dependence is not known. In this case, the problem of determining the dependence of the volumetric heat capacity and the thermal conductivity on the temperature based on experimental measurements of the temperature field arises.

Determining the thermal conductivity of a substance is an important problem, and it has been studied for a long time. This is confirmed by a large number of publications (e.g., see [1–9]). In these works, the inverse coefficient problems are studied theoretically, and numerical methods for their solution are developed. As for the simultaneous determination of the volumetric heat capacity and the thermal conductivity, there are very few works devoted to this issue.

In this paper, we consider one possible statement of such inverse coefficient problem. This statement is new. We have never seen the inverse coefficient problem in such formulation. The consideration is based on the Dirichlet problem for the one-dimensional unsteady-state heat equation. The inverse coefficient problem is reduced to a variational problem. The mean-root-square deviation of the calculated temperature field from its experimental value was chosen as the cost functional. An algorithm for the numerical solution of the inverse coefficient problem is proposed. It is based on the modern approach of Fast Automatic Differentiation (see [10,11]), which made it possible to solve a number of difficult optimal control problems for dynamic systems (e.g., see [12–14]). The examples of solving the inverse coefficient problem discussed in the paper confirm the efficiency of the proposed algorithm.

2   Formulation of the Problem

A layer of material of width L is considered. The temperature of this layer at the initial time is given. It is also known how the temperature on the boundary of this layer changes in time. The distribution of the temperature field at each instant of time is described by the following initial boundary value (mixed) problem:

$$C(T)\frac{\partial T(x,t)}{\partial t} - \frac{\partial}{\partial x}\left(K(T)\frac{\partial T(x,t)}{\partial x}\right) = 0, \qquad (x,t) \in Q, \qquad (1)$$

$$T(x,0) = w_0(x), \qquad\qquad\qquad 0 \leq x \leq L, \qquad (2)$$

$$T(0,t) = w_1(t), \qquad T(L,t) = w_2(t), \qquad 0 \leq t \leq \Theta. \qquad (3)$$

Here x is the Cartesian coordinate in the layer; t is time; $T(x,t)$ is the temperature of the material at the point with the coordinates x at time t; $Q = \{(0 < x < L) \times (0 < t < \Theta)\}$; $C(T)$ is the volumetric heat capacity of the material; $K(T)$ is the thermal conductivity of the material; $w_0(x)$ is the given temperature at the initial time $t = 0$; $w_1(t)$ is the given temperature on the left boundary of the layer; $w_2(t)$ is the given temperature on the right boundary of the layer.

If the dependence of the volumetric heat capacity $C(T)$ of a substance and its thermal conductivity $K(T)$ on the temperature T is known, then we can solve the mixed problem (1)–(3) to find the temperature distribution $T(x,t)$ in $\overline{Q}$. Below, problem (1)–(3) will be called the direct problem.

If the dependence of the volumetric heat capacity of a substance and its thermal conductivity on the temperature T is not known, it is of great interest to determine these dependencies. A possible statement of this problem (it is classified as an identification problem of the model parameters) is as follows: find the dependence of $C(T)$ and $K(T)$ on T under which the temperature field $T(x,t)$, obtained by solving the mixed problem (1)–(3), is close to the field $Y(x,t)$ obtained experimentally. The quantity

$$\Phi(C(T), K(T)) = \int\limits_{0}^{\Theta} \int\limits_{0}^{L} [T(x,t) - Y(x,t)]^2 \cdot \mu(x,t) dx\, dt \qquad (4)$$

can be used as the measure of difference between these functions. Here $\mu(x,t) \geq 0$ is a given weighting function. Thus, the optimal control problem is to find the optimal control $\{C(T), K(T)\}$ and the corresponding solution $T(x,t)$ of problem (1)–(3) that minimize functional (4).

The optimal control problems similar to this one are typically solved numerically using a descent method, which requires the gradient of functional (4) to be known.

3   The Optimal Control Problem in the Discrete Case

When solving the optimal control problem numerically the unknown functions $C(T)$ and $K(T)$, $T \in [a,b]$ were approximated by continuous piecewise linear functions (see [9]). The domain $Q = \{(x,t) : (0 < x < L) \times (0 < t < \Theta)\}$ was decomposed by the grid lines $\{\widetilde{x}_i\}_{i=0}^{I}$ and $\{\widetilde{t}^j\}_{j=0}^{J}$ into rectangles. At each node $(\widetilde{x}_i, \widetilde{t}^j)$ of Q characterized by the pair of indices (i,j) all the functions are determined by their values at the point $(\widetilde{x}_i, \widetilde{t}^j)$ (e.g., $T(\widetilde{x}_i, \widetilde{t}^j) = T_i^j$). In each rectangle, the thermal balance must be preserved. The result is a finite-difference scheme approximating the mixed problem (1)–(3). The resulting system of nonlinear algebraic equations was solved iteratively using the Gaussian elimination (see [15]). This approach was used to solve the mixed problem (1)–(3), and the function $T(x,t)$ (more precisely, its approximation T_i^j) was found.

To find the minimum of the cost function, the L-BFGS-B method [16] has been used. It is well known that it is very important for the gradient methods to determine the exact values of the gradients. In this work, the gradient of the cost function was calculated by the Adept application package [17], used in Reverse (Adjoint) mode.

It is possible that the use of the forward mode and the Levengberg-Marquard method, as in [18], would be more efficient, but in this work we did not study the method's performance.

4   Numerical Results

To test the performance of the proposed algorithm, several series of calculations were performed. All the results presented in this paper were based on the fact that the function

$$T_*(x,t) = \left(\frac{n+1}{\beta}\right)^{\frac{1}{n+1}} \cdot [\alpha \cdot (t + \gamma - x)]^{\frac{1}{m-n}}, \quad \alpha = \frac{m-n}{n+1}, \quad \beta = (n+1)^{\frac{m-n}{m+1}},$$

is the analytical solution to equation (1)

$$C(T)\frac{\partial T(x,t)}{\partial t} - \frac{\partial}{\partial x}\left(K(T)\frac{\partial T(x,t)}{\partial x}\right) = 0$$

for $\gamma = const$, $\quad C(T) = \beta \cdot T^n$, $\quad K(T) = T^m$.

In accordance with this, the following input data of the problem were used:

$$L = 1, \qquad \Theta = 1, \qquad \gamma = 1.5, \qquad Q = (0,1) \times (0,1),$$

$$w_0(x) = T_*(x,0) = \left(\frac{n+1}{\beta}\right)^{\frac{1}{n+1}} \cdot [\alpha \cdot (\gamma - x)]^{\frac{1}{m-n}}, \qquad (0 \le x \le 1),$$

$$w_1(t) = T_*(0,t) = \left(\frac{n+1}{\beta}\right)^{\frac{1}{n+1}} \cdot [\alpha \cdot (t + \gamma)]^{\frac{1}{m-n}}, \qquad (0 \le t \le 1), \quad (5)$$

$$w_2(t) = T_*(1,t) = \left(\frac{n+1}{\beta}\right)^{\frac{1}{n+1}} \cdot [\alpha \cdot (t + \gamma - 1)]^{\frac{1}{m-n}}, \qquad (0 \le t \le 1),$$

$$Y(x,t) = T_*(x,t), \qquad\qquad\qquad\qquad (x,t) \in Q.$$

For the solution of the direct problem, we used the uniform grid with the parameters $I = 40$ (the number of intervals along the axis x) and $J = 7000$ (the number of intervals along the axis t), which ensures the sufficient accuracy of computed temperatures and pulses. The segment $[a,b]$ was partitioned into 80 intervals ($N = 80$).

In the <u>first series</u> of calculations the problem of finding the volumetric heat capacity and the thermal conductivity with input data (5) at $m = 2$ and $n = 1$ was considered. In this case, it was assumed that in the cost functional (4) the weight function $\mu(x) \equiv 1$ ("field" functional). It should be immediately noted that the solution $\{C_*(T), K_*(T)\}$ of the formulated identification problem in the case of using the "field" functional is not unique. Indeed, if the solution to the inverse problem is found, then the functions $\{\lambda C_*(T), \lambda K_*(T)\}$ are also the solution to this inverse problem for any λ. Therefore, when solving the problem

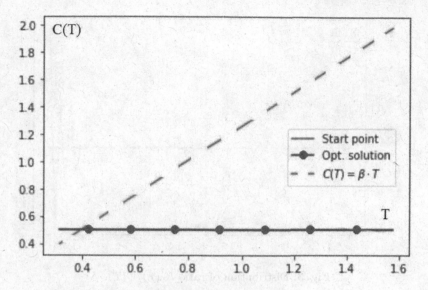

Fig. 1. Distribution of volumetric heat capacity.

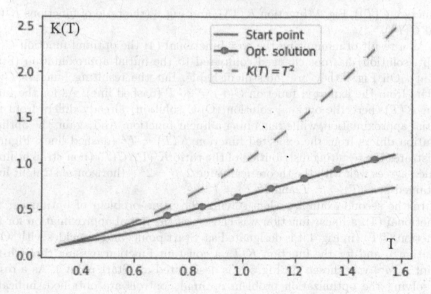

Fig. 2. Distribution of thermal conductivity.

of identifying the volumetric heat capacity and the thermal conductivity, an additional condition $K(a) = a^m$ was used.

In the first example, when solving the optimization problem for functional (4), the functions $\overline{C(T)} = 0.5$ and $K(T) = 0.5 \cdot T$ were chosen as the initial approximation. The results of solving the inverse problem are shown in Fig. 1

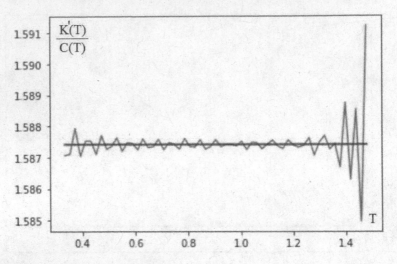

Fig. 3. Distribution of ratio $K'(T)/C(T)$.

(function $C(T)$), Fig. 2 (function $K(T)$), and Fig. 3 (the ratio of functions $K'(T)$ and $C(T)$).

As a result of minimizing the cost functional (4) the optimal function $C(T)$ (Opt. solution) has not changed compared to the initial approximation (Start point). On Fig. 1 they are indistinguishable, but the resulting function $C(T)$ differs from the expected function $C(T) = \beta \cdot T$ (dashed line). As for the function $K(T)$, here the optimal solution (Opt. solution) already differs from the initial approximation, while remaining a linear function. And again, the optimal solution differs from the expected function $K(T) = T^2$ (dashed line). Figure 3 illustrates the resulting distribution of the ratio $K'(T)/C(T)$ (non-straight line), which agrees well with the theoretical value $2/\beta = 2^{2/3}$ (horizontal straight line) obtained at $C(T) = \beta \cdot T$ and $K(T) = T^2$.

In the second example, when solving the same problem of optimizing the functional (4), a linear function was chosen as the initial approximation for the function $C(T)$ (in Fig. 4 it is designated as "Start point" and coincides with "Opt. solution"), and for the function $K(T)$ a constant function passing through the point (a, a^2) was chosen (in Fig. 5 it is designated as "Start point"). As a result of solving the optimization problem optimal controls are obtained, indicated in Fig. 4 and Fig. 5 as "Opt. solution". These figures show the following. The optimal function $C(T)$ does not differ from the initial approximation, and does not coincide with the expected function $C(T) = \beta \cdot T$ (dashed line in Fig. 4). The optimal function $K(T)$ is a parabola passing through the point (a, a^2), but also does not coincide with the expected function $K(T) = T^2$ (dashed line in Fig. 5). It is interesting that the ratio $K'(T)/C(T)$, as in example 1, remains almost constant over the temperature change interval $[a, b]$ and is in good agreement with the theoretical value $2/\beta = 2^{2/3}$ (horizontal straight line) obtained at $C(T) = \beta \cdot T$ and $K(T) = T^2$ (Fig. 6).

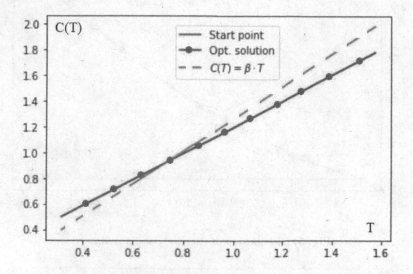

Fig. 4. Distribution of volumetric heat capacity.

Finally, in the third example of this series of calculations, when solving the optimization problem, the "expected" functions were chosen as initial approximations, namely, $C(T) = \beta \cdot T$, $K(T) = T^2$, $\beta = \sqrt[3]{2}$ (these functions were marked with dashed lines in the previous figures). As a result of optimization these functions have not changed much. Small changes, completely invisible on the graphs, are associated with the approximation of both the functions themselves and the differential equation. The maximum deviation of the ratio $K'(T)/C(T)$, built on the basis of optimal solutions, differs from the exact one by no more than 10^{-10}.

In the second series of calculations, the problem of finding the volumetric heat capacity and the thermal conductivity with input data (5) at $m = 4$ and $n = 2$ was considered. At the same time, as before, it was assumed that in the cost functional (4) the weight function $\mu \equiv 1$ ("field" functional). When solving the problem of identifying the volumetric heat capacity and the thermal conductivity, an additional condition $K(a) = a^m$ was used.

In the first example of this series of calculations, when solving the optimization problem, the "expected" functions were chosen as initial approximations, namely, $C(T) = \beta \cdot T^2$, $K(T) = T^4$, $\beta = \sqrt[5]{9}$ (these functions will be marked with dashed lines in the Figs. 7, 8 and 9). As a result of optimization these

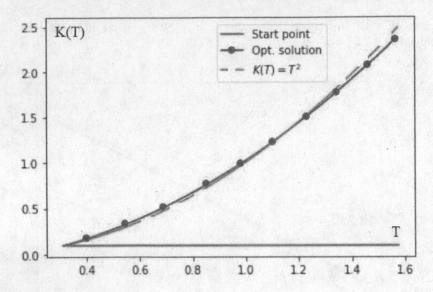

Fig. 5. Distribution of thermal conductivity.

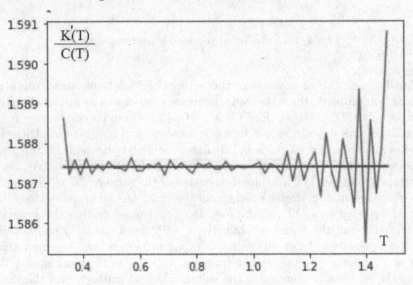

Fig. 6. Distribution of ratio $K'(T)/C(T)$.

functions have not changed much. Small changes, completely invisible on the graphs, are associated with the approximation of both the functions themselves and the differential equation. The maximum deviation of the ratio $K'(T)/C(T)$ constructed on the basis of optimal solutions differs from the exact one (dashed line in Fig. 7) by no more than 10^{-10}.

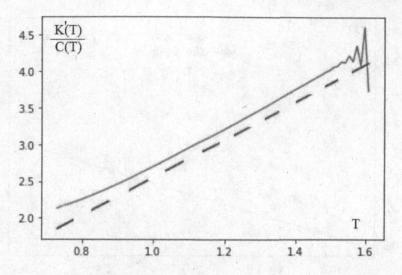

Fig. 7. Distribution of ratio $K'(T)/C(T)$.

In the second example of the second series of calculations, when solving the same problem of optimizing the functional (4), a linear function was chosen as the initial approximation for the function $C(T)$ (in Fig. 8 it is designated as "Start point" and coincides with "Opt. solution"), and for the function $K(T)$ a constant function passing through the point (a, a^4) was chosen (in Fig. 9 it is designated as "Start point"). As a result of solving the optimization problem optimal controls are obtained, indicated in Fig. 8 and Fig. 9 as "Opt. solution". These figures show the following. The optimal function $C(T)$ does not differ from the initial approximation, and does not coincide with the expected function $C(T) = \beta \cdot T^2$ (dashed line in Fig. 8). The optimal function $K(T)$ is a parabola passing through the point (a, a^4), but also does not coincide with the expected function $K(T) = T^4$ (dashed line in Fig. 9). In addition, in this example, the ratio $K'(T)/C(T)$ changes on a segment $[a, b]$ (solid line in Fig. 7), and it does not coincide with a similar line from the first example of the second series of calculations (dashed line in Fig. 7).

Finally, in the third example of the second series of calculations, when solving the optimization problem, the following functions were chosen as initial approximations, namely, $C(T) = (T - T(a)) \cdot (T(b) - T) + 1$ (see Fig. 10), $K(T) = T^4(a)$ (see Fig. 11) (these functions in Fig. 10 and Fig. 11 were highlighted with solid lines). As a result of optimization, the function $C(T)$ practically did not change (in Fig. 10, the initial approximation and the optimal solution are indistinguishable). As for the function $K(T)$, the optimal solution approximately has the form of a cubic parabola (see Fig. 11).

216   A. Albu et al.

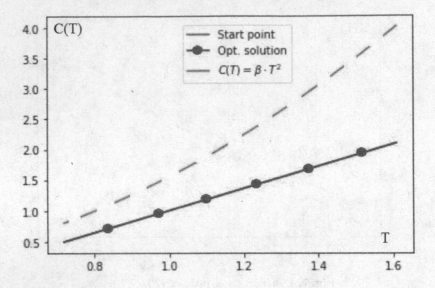

Fig. 8. Distribution of volumetric heat capacity.

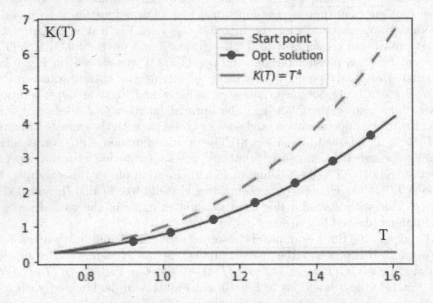

Fig. 9. Distribution of thermal conductivity.

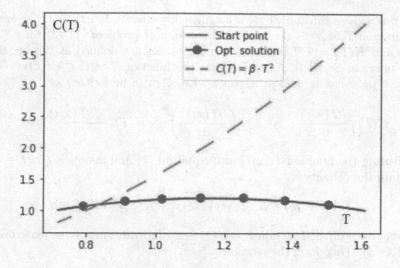

Fig. 10. Distribution of volumetric heat capacity.

5 Analysis of the Obtained Results

In all the examples considered in the previous section, the solution to the problem of identifying the volumetric heat capacity and the thermal conductivity turned out to be not unique. This is due to the specifics of the experimental temperature field used. Indeed, this field has the form

$$T_*(x,t) = V(t + \gamma - x), \tag{6}$$

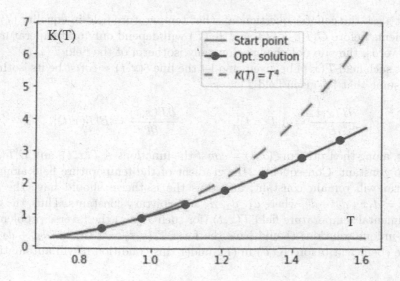

Fig. 11. Distribution of thermal conductivity.

where γ is a given constant, $V(\xi)$ is a sufficiently smooth function of it argument. The function $T_*(x,t)$ is the analytical solution of problem (1)–(3) for $C(T) = \beta \cdot T^n$ and $K(T) = T^m$ (all occurring quantities are defined at the beginning of the previous Section). Assuming that the function $T(x,t) \in C^2(Q) \cap C(\overline{Q})$, $C(T) \in C([a,b])$, and $K(T) \in C^1([a,b])$ Eq. (1) can be written as

$$C(T) \cdot \frac{\partial T(x,t)}{\partial t} - K'(T) \cdot \left(\frac{\partial T(x,t)}{\partial x}\right)^2 - K(T) \cdot \frac{\partial^2 T(x,t)}{\partial x^2} = 0. \qquad (7)$$

Substituting the function $T_*(x,t)$ into equation (7) and denoting $\xi = t + \gamma - x$ we obtain the equality

$$C(T_*) \cdot V'(\xi) - K'(T_*) \cdot (V'(\xi))^2 - K(T_*) \cdot V''(\xi) = 0. \qquad (8)$$

The level lines of the function $\xi = t + \gamma - x$ are isotherms of the experimental field $T_*(x,t)$. Therefore, the relation

$$C(T_*) \cdot V'(T_*) - K'(T_*) \cdot (V'(T_*))^2 - K(T_*) \cdot V''(T_*) = 0 \qquad (9)$$

is valid along an arbitrary isotherm of the field $T_*(x,t)$.

Thus, the unknown functions $C(T)$ and $K(T)$ on the segment $[a,b]$ must be connected by equality (9). Equality (9) shows that given arbitrarily $K(T) \in C^1([a,b])$, we can uniquely determine $C(T) \in C([a,b])$. And vice versa, by setting $C(T) \in C([a,b])$ and the value $K(T)$ at some point of the segment $[a,b]$, it can be uniquely determined $K(T) \in C^1([a,b])$ as a solution to the Cauchy problem. It was with such ambiguity that one had to deal with identification $C(T)$ and $K(T)$ in the previous examples. In relation (9), the coefficients before $C(T)$, $K(T)$, and $K'(T)$ depend only on T due to the specifics of the temperature field $T_*(x,t)$.

Let's ask the counter question: at what temperature field in equality (7), the coefficients before $C(T)$, $K(T)$, and $K'(T)$ will depend only on temperature (in other words, they do not change along any isotherm of the field $T(x,t)$)?

Let such field $T(x,t)$ be given, and let the line $\xi(x,t) = const$ be its isotherm. We assume that for given field $T(x,t)$

$$\frac{\partial T(x,t)}{\partial x} = A[T(x,t)], \qquad \frac{\partial T(x,t)}{\partial t} = B[T(x,t)],$$

that is, along the isotherm $\xi(x,t) = const$ the functions $A[T(x,t)]$ and $B[T(x,t)]$ remain constant. Consequently, the gradient of the temperature field along the isotherm will remain constant, and thus the isotherm should have the form $\xi(x,t) = d_1 x + d_2 t + d_3$, where d_1, d_2, d_3 are arbitrary constants. Thus, the given experimental temperature field $T(x,t)$ (for such $T(x,t)$ the inverse problem has a non-unique solution) should have the form $T_*(x,t) = V(d_1 x + d_2 t + d_3)$. As for the coefficient before $K(T)$ in (7), under this condition it will automatically depend only on the temperature:

$$\frac{\partial^2 T(x,t)}{\partial x^2} = \frac{\partial}{\partial x} \left(\frac{\partial T(x,t)}{\partial x} \right) = \frac{\partial}{\partial x} \left(A[T(x,t)] \right)$$

$$= A'(T) \cdot \frac{\partial T(x,t)}{\partial t} = A'(T) \cdot A(T) = D(T).$$

6   Conclusion

The formulated inverse problem always has a non-unique solution. Therefore, you must always specify one additional condition to isolate a single solution. However, there are experimental temperature fields at which setting one additional condition is not enough. In this case, it is necessary to fix either the thermal conductivity or the volumetric heat capacity.

References

1. Kozdoba, L.A., Krukovskii, P.G.: Methods for Solving Inverse Thermal Transfer Problems. Naukova Dumka, Kiev (1982). (in Russian)
2. Alifanov, O.M.: Inverse Heat Transfer Problems. Springer, Heidelberg (2011). https://doi.org/10.1007/978-3-642-76436-3
3. Vabishchevich, P.N., Denisenko, A.Yu.: Numerical methods for solving inverse coefficient problems. Method of Mathematical Simulation and Computational Diagnostics, Mosk. Gos. Univ., Moscow, pp. 35–45 (1990). (in Russian)
4. Samarskii, A.A., Vabishchevich, P.N.: Difference methods for solving inverse problems of mathematical physics. Fundamentals of Mathematical Simulation, Nauka, Moscow, pp. 5–97 (1997). (in Russian)
5. Samarskii, A.A., Vabishchevich, P.N.: Computational Heat Transfer. Editorial URSS, Moscow (2003). [in Russian]
6. Marchuk, G.I.: Adjoint Equation and the Analysis of Complex Systems. Nauka, Moscow (1992). (in Russian)
7. Kabanikhin, S.I.: Inverse and ill-posed problems. Sib. Elektron. Mat. Izv. **7**, 380–394 (2010)
8. Kabanikhin, S.I.: Inverse problems of natural science. Comput. Math. Math. Phys. **60**(6), 911–914 (2020). https://doi.org/10.1134/S0965542520060044
9. Albu, A.F., Zubov, V.I.: Identification of thermal conductivity coefficient using a given temperature field. Comput. Math. Math. Phys. **58**(10), 1585–1599 (2018)
10. Evtushenko, Y.G.: Computation of exact gradients in distributed dynamic systems. Optim. Methods Software **9**, 45–75 (1998). https://doi.org/10.1080/10556789808805686
11. Evtushenko, Y.G.: Optimization and Fast Automatic Differentiation. Vychisl. Tsentr Ross. Akad, Nauk, Moscow (2013). (in Russian)
12. Evtushenko, Y.G., Zasukhina, E.S., Zubov, V.I.: Numerical optimization of solutions to Burgers' problems by means of boundary conditions. Comput. Math. Math. Phys. **37**, 1406–1414 (1997)
13. Albu, A.F., Zubov, V.I.: Investigation of the optimal control of metal solidification for a complex-geometry object in a new formulation. Comput. Math. Math. Phys. **54**(12), 1804–1816 (2014). https://doi.org/10.1134/S0965542514120057

14. Albu, A.F., Zubov, V.I.: On the efficient solution of optimal control problems using the fast automatic differentiation approach. Tr. Inst. Mat. Mekh. UrO Ross. Akad. Nauk. **21**(4), 20–29 (2015)
15. Samarskii, A.A.: Theory of Finite Difference Schemes. Marcel Dekker, New York (2001)
16. Qiu, Y.: L-BFGS++ (2021). https://github.com/yixuan/LBFGSpp/
17. Hogan, R.J.: Fast reverse-mode automatic differentiation using expression templates in C++. ACM Trans. Math. Softw. (TOMS) **40**(4), 26–42 (2014)
18. Gorchakov, A.Y.: About software packages for fast automatic differentiation. Informacionnye tekhnologii i vychislitel'nye sistemy **1**, 30–36 (2018)

Application of the Interpolation Approach for Approximating Single-Machine Scheduling Problem with an Unknown Objective Function

Alexander Lazarev[1] , Egor Barashov[1]([✉]) , Darya Lemtyuzhnikova[1,2] , and Andrey Tyunyatkin[1]

[1] Institute of Control Sciences, 65 Profsoyuznaya Street, Moscow 117997, Russia
barashov.eb@gmail.com
[2] Moscow Aviation Institute, 4 Volokolamskoe Highway, Moscow 125993, Russia

Abstract. In this paper we consider an approximation-interpolation approach based on the combination of interpolation method and approximation method. An approximation method for single-machine scheduling theory problem with an unknown objective function that depends on the completion times of jobs is studied. The idea of approximating an unknown function by a linear function with some weight coefficients is considered. An approximation-interpolation algorithm has been developed, which allows to determine the values of weight coefficients of the objective function of single-machine problem of the scheduling theory. In this algorithm the minimization of the total weighted completion times according to the given set of values of the problem parameters and the corresponding known optimal schedules was carried out. Experiments with Lagrange polynomials interpolation and cubic splines have been conducted. The hypothesis of the necessity to take into account long-range bounds when calculating weighting coefficients has been confirmed.

Keywords: Scheduling · Approximation · Interpolation

1 Introduction

Scheduling problems constantly arise both in the planning of production processes and in everyday life. It is a set of works that are characterized by release time, processing time and due date. The purpose of such problems is to minimize or maximize some objective function, which represents some optimal criterion of the problem. Special consideration should be given to problems whose objective function is not defined. This paper presents an approximation-interpolation

The results of the sections 1–3 were obtained within the State Assignment of the Ministry of Science and Higher Education of the Russian Federation (theme No. 122041300137-2). The results of the section 5–6 were obtained within the RFBR and MOST grant (project No. 20-58-S52006). The results of the section 7 were obtained within the RSF grant (project No. 22-71-10131).

method that allows to rebuild the objective function from some known solutions to the initial problem. A linear function is used as an approximate function. That is, a problem $1|r_j| \sum w_j C_j$ is considered that approximates some problem for a single machine with an unknown objective function.

General case of the problem $1|r_j| \sum w_j C_j$ with weight coefficients w_j is NP-hard in the strong sense so a polynomial algorithm for the solution is unlikely to exist. For the single-machine problem, the proof can be found in [1]. Such problems can be solved either with an exact or approximate approach. The approximation method allows us to obtain for the NP-hard problem a fast enough solution as far as the objective function is concerned. In particular, it works for problems where the objective function is not predefined [2]. The approximation method consists in the fact that after reducing the initial problem to the solution of a system of linear inequalities with respect to the unknown values of the weight coefficients, they can be found by approximating the found boundary values.

There exist many different applications of approximation for a wide range of problems. Here is a brief review of the current literature of recent years, in which the approximation approach is used. For instance, in [3] the method of saddle point approximation is introduced to perform reliability analysis in design optimization for engineering structures design problem. Approximated methods of reliability analysis are widely used as an effective alternative to the exact reliability assessment to improve the efficiency of calculations. The main idea is to evaluate the reliability of the corresponding approximated function after approximating the corresponding limit state function. However, both of the above methods must convert random variables with an arbitrary distribution into random variables with a standard normal distribution. When the random variables involved follow a particular non-normal distribution, this conversion is a nonlinear process. This will increase the nonlinearity of the initial problems. To solve this problem, this study uses the saddle point approximation method. Saddle point approximation is a powerful strategy for approximating the distribution of random variables, in other words, accurately approximating the distribution function of random variables.

There is no universal way to approximate the unknown function, so various algorithms are still being developed that provide more accurate approximation coefficients than the best known classical algorithms. For example, the quantum approximate optimization algorithm is used for this purpose in [4]. This is a heuristic method for solving combinatorial optimization problems on quantum short-range computers. This work uses the recently proposed digital-analog paradigm for quantum computing, in which the versatility of programmable quantum computers and the error tolerance of quantum simulators are combined to improve platforms for quantum computing.

In [5], the ship performance model approximates the fuel consumption by calculating the ship resistance, i.e., the required propeller thrust to move the ship at a given speed, using real weather data as input to the model. The total resistance is modeled as the sum of the resistance to calm water, the resistance caused by waves, the resistance caused by sea currents, and the resistance to air.

Once the drag is known, the fuel consumption is estimated considering the fixed hull, shaft and propeller efficiency and the standard specific fuel consumption curve for marine engines.

The interpolation approach [6,7], which is used to speed up the work of already existing algorithms for solving scheduling problems, is planned to be applied to this problem. In conjunction with the approach under consideration, algorithms for solving scheduling problems can be used, as discussed, for example, in [6,8,9]. The approximation algorithm described above can be used in conjunction with the approach under consideration.

The structure of this paper is organized as follows. In Sect. 2 the mathematical formulation of the problem $1||\sum w_j C_j$ will be given; in Sect. 3 a method for solving the problem is proposed; in Sect. 4 a numerical study of the constructed approximation algorithm is carried out and its final complexity is calculated. The Sect. 5 presents the main theorem on which the approximation algorithm and the pseudocode of this algorithm are based. The Sect. 6 presents the main theoretical aspects of the approximation approach in connection with interpolation: a modified theorem is presented, a pseudocode is formulated, a basic experiment is given to check the sorting condition of the interpolation nodes in non-decreasing order. The Sect. 7 presents several experiments that illustrate the work of the approximation-interpolation approach, and the corresponding conclusions are drawn.

2   Formulation of the Problem

In this paper we investigate the scheduling problem $1|r_j|\sum F_j(C_j)$. On a single machine it is needed to process n jobs. For each job $j \in J$, $J = \{1, 2, \ldots, n\}$, the release time of the job r_j and the processing times p_j are given. Let L be the number of given instances of the initial problem. There is no precedence relationship and preemptions of the jobs are not allowed. At any moment, the machine can process no more than one job. A schedule π is a permutation of n jobs $(j_1, \ldots, j_n)$. We need to find a schedule π^0 in which the total of unknown objective function $F_j(C_j(\pi))$ is minimized, where $C_j(\pi)$—the completion time of the job j, $j \in J$. Then we use the idea of approximating the unknown function by the function of the total weighted completion times $\sum w_j C_j(\pi)$, since we cannot approximate to a function of unknown form, we will for certainty take the way as in [2], where $w_j > 0$—the unknown weighting coefficient of the corresponding job j. Then the completion times of the jobs at the schedule $\pi = (j_1, \ldots, j_n)$ are defined as follows:

$$C_{j_1}(\pi) = r_{j_1} + p_{j_1}; C_{j_k}(\pi) = \max\{C_{j_{k-1}}(\pi), r_{j_k}\} + p_{j_k}, k = 2, \ldots, n. \quad (1)$$

The problem $1|r_j|\sum w_j C_j$ is NP-hard [1]. In addition we don't know the weights of $w_j, j \in J$. We propose to solve the problem of approximation of weight coefficients w_j. By approximation we mean the construction of approximate values of weight coefficients w_j using the known parameter sets of instances of the problem $1|r_j|\sum w_j C_j$. Lets formulate vectors r and p as follows: $r =$

$\{r_1,\ldots,r_n\}, p = \{p_1,\ldots,p_n\}$. Lets proceed to the definitions of the instance problem $1|r_j|\sum w_j C_j$.

Definition 1. *Let the scheduling problem* $\mathfrak{P}(1|r_j|\sum w_j C_j, J, w, r, p, \pi)$ *be defined as follows:*

$$\mathfrak{P}(1|r_j|\sum w_j C_j, J, w, r, p, \pi) = \sum_{k=1}^{n} w_{j_k} C_{j_k}(\pi), \qquad (2)$$

where $C_{j_k}(\pi)$ *is according to (1),* π *is the schedule (permutation) from* $n!$ *schedules. For any parameters* r *and* p *that form a solvable case* $\mathfrak{P}(1|r_j|\sum w_j C_j, J, w, r, p, \pi)$, *we call it an instance problem* $1|r_j|\sum w_j C_j$ *for any permutation* π.

Each instance problem is uniquely defined by a set of parameters of the instance problem.

Definition 2. *I is the set of instances defined by* (J, p, r, π). *For any instance* I *the set of values* $\{(r^I = \{r_1,\ldots,r_n\}, p^I = \{p_1,\ldots,p_n\})\}$ *and schedule* $\pi = \{j_1,\ldots,j_n\}$ *will be called the parameters of the instance problem* $\mathfrak{P}(1|r_j|\sum w_j C_j, J, w, r, p, \pi)$, *where* r^I, p^I *means that values* r_j *and* $p_j, j \in J$, *and* π *is one of* $n!$ *schedules, belong to the instance* I. *The value* $C_j^l(\pi)$ *is denoted as a completion time of the job* j *in schedule* π *from instance* l.

We don't know $F_j(C_j)$ and would like to approximate this initial function by $w_j C_j, j \in J$, but we do not know the values of the weight coefficients.

Definition 3. *Let L instances of the problem* $\mathfrak{P}(1|r_j|\sum w_j C_j, J, w, r, p, \pi)$ *be given, for each of which the corresponding optimal schedules* $\pi_l^0, l = 1,\ldots,L$, *are known. Lets call the approximation problem of weight coefficients of the objective function by PAWCOF.*

3 Solvability for Approximation of the Problem $1|r_j|\sum w_j C_j$ in Polynomial Time

To solve the *PAWCOF*, it is necessary to minimize the total of weighted completion times of jobs with unknown weight coefficients, but a known set of optimal schedules is given. The method for solving the *PAWCOF* is based on determining the optimality of schedules π_l^0 for all solvable instances $l = 1, 2,\ldots,L$. That is, the completion time on the right side of the inequality below refers to the optimal schedule π_l^0, and the completion time on the left side refers to some unknown schedule π. Consequently, it follows from the optimality of the value of the objective function considered on the right that its value will either be exactly equal to the left part, or will be less:

$$\sum_{j=1}^{n} C_j^l(\pi)w_j \geq \sum_{j=1}^{n} C_j^l(\pi_l^0)w_j.$$

Note that this type of objective function allows to move the completion times to the left side of the inequality, which allows to formulate the following system of inequalities:

$$\sum_{j=1}^{n}\Big(C_j^l(\pi) - C_j^l(\pi_l^0)\Big)w_j \geq 0, \quad \forall\pi,\ l \in 1,\dots,L. \qquad (3)$$

It follows from the system of inequalities (3) that the values of weight coefficients w_j are generally determined by a system of $L(n!-1)$ inequalities, where n—the number of jobs of the problem $1|r_j|\sum w_j C_j$, and L—the number of solvable instances of the problem. The dependence of the number of inequalities in the system on the number of jobs is not polynomial. The question arises: in what cases is it possible to reduce the complexity of the problem to polynomial? This is possible when among $L(n!-1)$ inequalities of the system there is some constant M of independent inequalities. And the other inequalities of the system (3) will depend on them.

Lets consider the simple case of the $PAWCOF$ for $1|r_j|\sum w_j C_j$. Let the system (3) contain only one inequality, that is, a single instance is given, $L = 1$. For n jobs release times r_j and processing times p_j are given, and the optimal schedule π^0 is known. Then for an arbitrary schedule π the inequality is valid:

$$\sum_{j=1}^{n}\Big(C_j(\pi) - C_j(\pi^0)\Big)w_j \geq 0. \qquad (4)$$

Consider a subset of this case for which a subsystem of M inequalities can be selected such that the solution of this subsystem coincides with the solution of the initial system, and the number M does not exceed n.

Since each of the strict inequalities of the form (4) corresponds to some non-optimal schedule π, so among $(n!-1)$ non-optimal schedules there are M schedules $\Pi = \{\pi_1,\dots,\pi_M\}$. The schedule π_i, $i = 1,\dots,M$, contribute to the fact that the solution of the corresponding system of inequalities of the form (4) coincides with the solution of the initial system, and the number M depends polynomially on n. That is, for any schedule $\pi \notin \Pi$, an inequality of the form (4) corresponding to schedule π is a consequence of the system of inequalities corresponding to schedule π_i, $i = 1,\dots,M$. Let us summarize this conclusion as the definition of the solvability of the $PAWCOF$ in polynomial time.

Lemma 1. *The PAWCOF (3) will be solvable in polynomial time from the problem dimension if the subsystem of m inequalities (4) can be selected so that the solution of this subsystem coincides with the solution of the initial system, and the number of m doesn't exceed the number of n.*

The proof of the lemma is given in [10].

4 Transition to the Special Case of the Problem $1||\sum w_j C_j$

The problem $1|r_j|\sum w_j C_j$ with known weight coefficients w_j is NP-hard in the strong sense [1]. Lets consider the definition of the special case for this problem.

Definition 4. *An instance of a problem $1|r_j|\sum w_j C_j$ will be called a special case if the corresponding parameters of this problem have some formalized regularity. If there exists some algorithm that allows to obtain the exact solution of a particular case of the problem in polynomial or pseudo-polynomial time, such a particular case will be called polynomially or pseudo-polynomially solvable case of the problem.*

The problem $1||\sum w_j C_j$ is a polynomially solvable case of $1|r_j|\sum w_j C_j$ according to the generalized Smith theorem [11]. Next, we will consider the problem of approximating the weight coefficients of the problem $1||\sum w_j C_j$.

Remark 1. For the polynomially solvable case of the problem $1||\sum w_j C_j$ the definition of the *PAWCOF* (3) is also correct.

Lets reformulate the definition of the parameters of the instance problem for the special case $1||\sum w_j C_j$.

Definition 5. *The set of values $\{(p_j^I)\} = \{p_1^I, \dots, p_n^I\}$ will be called the parameters of the instance problem $1||\sum w_j C_j$, where p_j^I means that this value p_j belongs to instance I.*

The transition to the special case of the problem is associated with the fact that some formalized regularity of the special case can be used to construct instances of the problem on which the approximation is built. In the problem $1||\sum w_j C_j$ a formalized regularity of the following kind is established. The simultaneous release times of jobs on the machine is assumed: $r_1 = \dots = r_n = r$. In this case, the optimal schedule will be the schedule made in the order of non-decreasing values $p_j/w_j, j \in J$. That is, for the parameters of the instance $p_j^I = \{p_1^I, \dots, p_n^I\}$ problem $1||\sum w_j C_j$, there exists an optimal schedule $\pi^* = (j_1, j_2, \dots, j_n)$ according to [11], such that

$$\frac{p_{j_1}^I}{w_{j_1}} \le \frac{p_{j_2}^I}{w_{j_2}} \le \dots \le \frac{p_{j_n}^I}{w_{j_n}}, I = 1, \dots, L. \tag{5}$$

An inequality of the form (5) is valid for all L given instances with parameters $\{p_1^l, \dots, p_n^l\}$ of the problem and corresponding optimal schedules $\pi_l^0 = (j_1^l, j_2^l, \dots j_n^l), l \in \{1, \dots, L\}$. Next, to determine the polynomially solvable *PAWCOF*, we turn to the definition of the effective system of inequalities formulated in [2]. This notion is a formalized definition of the polynomially solvable *PAWCOF* for $1||\sum w_j C_j$.

Definition 6. *A system of inequalities where jobs i, j were chosen arbitrarily among the total number of jobs,*

$$X(i,j) \leq \frac{w_j}{w_i} \leq Y(i,j), \qquad (6)$$

where:

$$X(i,j) = \max_{k \in K_{j,i}} \left(\frac{p_j^k}{p_i^k} \right),$$

$$Y(i,j) = \min_{k \in K_{i,j}} \left(\frac{p_j^k}{p_i^k} \right),$$

$$K_{i,j} = \{k \in K \; : \; \pi_k^0 = (\ldots, i, \ldots, j, \ldots)\},$$

$$K_{j,i} = \{k \in K \; : \; \pi_k^0 = (\ldots, j, \ldots, i, \ldots)\},$$

$$i, j \in J, \quad i \neq j,$$

*will be called **effective system** of inequalities of the weight coefficient approximation problem PAWCOF for the case $r_1 = \ldots = r_n$.*

$$X(i,j) := \max\Big\{ X(i,j); \max_{l=1,\ldots,n, l \neq i, l \neq j} \{X(i,l)X(l,j)\} \Big\}, \quad i, j \in J, \; i \neq j. \quad (7)$$

The calculation of matrices $\tilde{X}, \tilde{Y}$ is the most time-consuming part of the algorithm in terms of computational complexity. Indeed, for each pair i, j involves $(n-2)$ calculations of the product $X(i,l)X(l,j)$ and finding the maximum of $(n-1)$ values. Thus, the number of operations required for a single iteration of the procedure (7) is

$$O\big((n-2) + (n-1)\big) = O(n).$$

To find matrices $\tilde{X}, \tilde{Y}$ it is necessary to repeat this procedure (7) for all possible pairs of requirements i, j until after another iteration no such pair i, j is found that element $X(i,j)$ can be increased. However, if we take advantage of the property that after each iteration of the procedure the element $X(i,j)$ either increases or remains unchanged, the number of iterations of the procedure can be significantly reduced by specifically choosing the order of consideration of pairs i, j so that the procedure (7) will not be executed for each pair of requirements more than twice. In this case, the number of operations required to compute the matrices $\tilde{X}, \tilde{Y}$, is

$$O\Big(2 \cdot \frac{n(n-1)}{2} \cdot n\Big) = O(n^3).$$

5   Approximation Algorithm for Solving the *PAWCOF*

In order to find a solution to the *PAWCOF*, it is necessary to find a solution to an effective system of inequalities. At "input" there is a system $\frac{n(n-1)}{2}$ of dual inequalities of the form:

$$X(i,j) \leq \frac{w_j}{w_i} \leq Y(i,j), \quad i, j \in J, \; i \neq j. \qquad (8)$$

The system (9) can also be reduced to a system of linear homogeneous inequalities by writing it as $n(n-1)$ simple linear inequalities:

$$\begin{cases} w_j - Y(i,j)w_i \le 0, \\ -w_j + X(i,j)w_i \le 0. \end{cases}$$

For such systems there are many algorithms to find a system of formative vectors sufficient to write the general formula for non-negative solutions. However, the resulting system has a special form that differs significantly from the general one. Let us give further a theorem, where the solution of an effective system of inequalities (6) is formulated explicitly.

Definition 7. *The iterative application of the formulas:*

$$X(i,j) := \max\Big\{ X(i,j); \max_{l=1,\dots,n, l \ne i, l \ne j} \{ X(i,l)X(l,j) \} \Big\} = \frac{1}{Y(j,i)}, \qquad (9)$$

where $i,j \in J$, $i \ne j$, yields the amplified matrices $\tilde{X}, \tilde{Y}$, which are matrices whose values are extreme estimates of the lower and upper bounds, respectively.

Next, the amplified matrices will be used to find the values of the unknown weight coefficients. Any value from the interval $[\tilde{X}(l,j), \tilde{Y}(l,j)]$, where $j,l \in J, j \ne l$, is a solution to the initial problem. We take the middle of this interval for certainty.

The procedure (9) must be repeated for all possible pairs i,j. It is necessary to go through all possible i,j more than once: it is necessary to repeat the procedure (9) until, as a result of the next cycle, none of the inequalities of the effective system can be strengthened. After a sufficient number of repetitions of the procedure (9) we denote matrices X, Y by $\tilde{X}, \tilde{Y}$.

Thus it can be stated that if for any pair i,j: $w_j/w_i = Z \in [\tilde{X}(i,j); \tilde{Y}(i,j)]$, then there is no contradictory pair among the remaining system inequalities after appropriate simplifications. Hence, for any $Z \in [\tilde{X}(i,j); \tilde{Y}(i,j)]$ there exists a solution to the system (5) of inequalities such that $w_j/w_i = Z, j,i \in J, j \ne i$.

Suppose we don't have an interior point among the sets of bounds for the weighting coefficients. In this case $\tilde{X}(i,j) = \tilde{Y}(i,j) \; \forall i,j, \; i \ne j$. Then using the equality $\tilde{X} = 1/\tilde{Y}$ we obtain that this problem turns into the case when $p = const, w = const$. This trivial case is solved by setting the jobs by the non-decreasing order of deadlines.

The weight coefficients scale: the problems $1||\sum w_j C_j$ and $1||\sum (\gamma w_j)C_j$ are equivalent $\forall \gamma > 0$, and if the set of coefficients $w = \{w_1, \dots, w_n\}$ is a solution of the system of inequalities (initial or effective), then the set $\gamma w = \{\gamma w_1, \dots, \gamma w_n\}$ is a solution of this system. This fact makes it possible to arbitrarily choose a value of one of the required weight coefficients, solve the system with respect to the other weight coefficients, and scale the resulting solution without losing the solution of the initial system of inequalities.

Assume, for example, $w_1 = 1$. For arbitrary $j \in \{2, \dots, n\}$ we have:

$$\tilde{X}(1,j) \le w_j \le \tilde{Y}(1,j). \qquad (10)$$

The inequalities (10) describe a parallelepiped in a hyperspace of dimension $(n-1)$. Moreover, for all w_j and for any $Z \in [\tilde{X}(1,j); \tilde{Y}(1,j)]$ there exists a solution of the system such that $w_j/w_1 = Z$ or $w_j = Z, j = 2, \ldots, n$. In particular, when $w_1 = 1$ there exists a solution such that $w_j = \tilde{X}(1,j)$ and such that $w_j = \tilde{Y}(1,j)$. In other words, the section of the set of solutions by the plane $w_1 = 1$ lies inside the parallelepiped described by inequalities (10) and has at least one common point with each of its edges. Hence since the center of this parallelepiped is an interior point of the solution set of the initial system of inequalities, it is also a solution of the initial system.

Thus the following was proved

Theorem 1. *Vector* $w = (w_1, \ldots, w_j, \ldots, w_n)$, *where*

$$w_j = \begin{cases} 1, & j = l; \\ (\tilde{X}(l,j) + \tilde{Y}(l,j))/2, & j \neq l. \end{cases}$$

is the initial solution of the effective system of inequalities (the index l is arbitrarily defined).

Since the *PAWCOF* problem does not aim to find all possible solutions of the system, below we will propose a fast algorithm for finding a partial solution of the system of inequalities of the obtained form from (6). The algorithm for approximating the weight coefficients of the objective function is based on solving an effective system of inequalities (6).

To approximate the coefficients of $w_j, j \in J$ it is necessary to:

1. Construct sets $K_{i,j}, K_{j,i}, \forall i \neq j$;
2. Fill in the matrices $X(i,j), Y(i,j)$;
3. Calculate the matrices $\tilde{X}(i,j), \tilde{Y}(i,j)$;
4. Calculate $w_j = \begin{cases} 1, & j = l \\ (\tilde{X}(l,j) + \tilde{Y}(l,j))/2, & j \neq l \end{cases}$, where the index l is chosen arbitrarily.

The implementation of this algorithm is presented in pseudocode in Appendix.

6 Modification of the Approximation Algorithm for Solving the *PAWCOF* Using Interpolation

In the conditions when only a relatively small dataset is available, it is not possible to construct the bounds of $X(s,j)$ and $Y(s,j), s, j \in J$ that are close enough to each other to adequately estimate the values of the weight coefficients w_j. Therefore, to increase the accuracy of the approximation, additional information can be used, "hidden" in the relative location of all boundaries. Thus, with a small number N of instances of I_k, the interpolation modification of the approximation approach should differ from its standard implementation by a smaller error.

Now we shall prove a theorem on the basis of which an approximation-interpolation algorithm will be constructed.

Theorem 2. *Let* $\xi(k,j) = X(k+1,j)$ *while* $k < n$ *and* $\xi(s,j) = Y(k+1-n,j)$ *with* $k \geq n$. *If the following values* $\xi(k,j)$ *not sorted in ascending order, we will change the numbering of the instances so that the following condition is met:*

$$\xi(k_2,j) \geq \xi(k_1,j), \ \forall k_1, \forall k_2, \ k_2 > k_1.$$

This is possible due to the arbitrariness of the numbering of instances. Then the weight coefficient w_j *at* $j \neq s$ *can be estimated using the following formula:*

$$w_j = \sum_{k=0}^{2n-1} \frac{\prod_{i=0\ldots(2n-1), i\neq k}(n - \frac{1}{2} - i)}{\prod_{m=0\ldots(2n-1), m\neq k}(k - m)} \xi(k,j). \tag{11}$$

Proof. To approximate the weight w_j it is necessary to calculate the value $\tilde{w}_j$, lying between $\max_k X(k,j)$ and $\min_k Y(k,j)$.

What is more:

1. The instances are numbered in such a way that $\xi(k,j)$ does not decrease with the growth of the number k (by the condition of the theorem).
2. X and Y are matrices of size $n \times n$.

Then it is possible to obtain the following equations:

$$\xi(n-1,j) = X(n,j) = \max_k X(k,j), \tag{12}$$

$$\xi(n,j) = Y(1,j) = \min_k Y(k,j), \tag{13}$$

thus, the value $\tilde{w}_j$ lies between $\xi(n-1,j)$ and $\xi(n,j)$.

The interpolation curve is built using the points $(0, \xi(0,j)), \ldots, (2n-1, \xi(2n-1,j))$ and its value is calculated at the point $(n - 1/2, \xi(n - 1/2, j))$ which is located between $\xi(n-1,j)$ and $\xi(n,j)$.

To calculate the interpolation curve, Lagrange interpolation formula is used:

$$\mathbf{L}_m(x) = \sum_{k=0}^{m} \frac{\prod_{i\neq k}(x - x_i)}{\prod_{j\neq k}(x_k - x_j)} f(x_k). \tag{14}$$

The interpolation nodes that are used are $(0, \xi(0,j)) \ldots (2n-1, \xi(2n-1,j))$ and then Largange polynome value is calculated at the point $(n - 1/2, \xi(n - 1/2, j))$. Then the following formula is obtained, which completes the proof of the theorem:

$$\tilde{w}_j = \mathbf{L}_m\left(n - \frac{1}{2}\right) = \sum_{k=0}^{2n-1} \frac{\prod\limits_{i=0...(2n-1),i\neq k}\left(n - \frac{1}{2} - i\right)}{\prod\limits_{m=0...(2n-1),m\neq k}(k-m)}\xi(k,j). \tag{15}$$

The modification of the approximation algorithm is formulated as follows.

1. Obtain the sets $K_{i,j}$, $K_{j,i}$;
2. Fill the matrices $X(i,j)$, $Y(i,j)$;
3. Calculate the matrices $\breve{X}(i,j)$, $\breve{Y}(i,j)$;
4. Calculate the value $w_j = \sum_{k=0}^{2n-1} \frac{\prod\limits_{i=0...(2n-1),i\neq k}\left(n - \frac{1}{2} - i\right)}{\prod\limits_{m=0...(2n-1),m\neq k}(k-m)}\xi(k,j)$, where $\xi(k,j)$ is

   calculated using the Theorem 2.

The implementation of this algorithm is presented in pseudocode in Appendix without taking into account the recalculation of data at the last step. After executing the main algorithm, w_j recalculated taking into account the interpolation. The pseudocode for recalculation using cubic interpolation is not given, since this is the standard algorithm.

The condition of the Theorem 2 requires sorting the interpolation nodes in non-decreasing order. Thus, the values of the interpolation polynomial should not decrease.

In practice, when implementing the algorithm, it makes sense to check this condition, since the Lagrange interpolation polynomial tends to form "bursts" if the number of the interpolation nodes is too high.

An instance of such case was given in [7] and is illustrated on the graph below (Fig. 1).

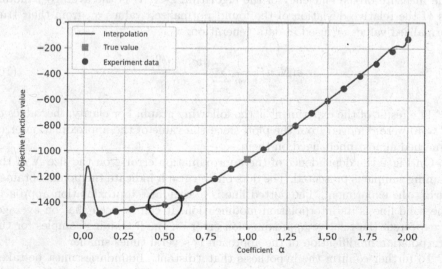

Fig. 1. "Bursts" on the borders of the interpolation interval

It should be noted that in the course of numerical experiments conducted by the authors, the indicated condition has never been violated. However, it is not possible to guarantee the fulfillment of this condition - especially with large training samples. Therefore, when implementing the approximation-interpolation algorithm, it is also recommended to implement the test mentioned above.

7   Data Generation and Numeric Experiments

Let's move on to the description of data generation. To conduct the numeric experiments, 100 sets of 100 instances were generated in each, with each instance consisting of 10 jobs. For each experiment n of weight coefficients $w_j^0, j \in J$ and L instances $I_k = \{p_1^k, \ldots, p_n^k\}, k = 1, \ldots, L$ were generated. All quantities have an even distribution on the interval $[0; 1])$. Then the constructed algorithm was run for the generated data, the result of which is a set of found values of weight coefficients $w_j, j \in J$. The number of instances is limited by the polynomial computational complexity of the problem.

Then the relative approximation error of ε was calculated. To compare the found w_j and the true w_j^0 values of the weighting coefficients, both sets are normalized (scaling is allowed due to the linearity of the objective function):

$$||w|| = \sqrt{\sum_{j=1}^n w_j^2},$$

$$w_j = \frac{w_j}{||w||}, \quad w_j^0 = \frac{w_j^0}{||w^0||}.$$

The measure of the efficiency of the algorithm $\varepsilon(N, n)$ is the average j modulus of the relative deviation of the found normalized values w_j from their true normalized values w_j^0 used in data generation:

$$\varepsilon(N, n) = \frac{1}{n} \sum_{j=1}^n \frac{|w_j - w_j^0|}{w_j^0}. \tag{16}$$

The result of the experiment is the following graph. For clarity, the values of $N > 40$ were removed from the plot, since the value of the approximation error remained almost unchanged on them.

On Fig. 2 the dependence of the approximation error ε on the size N of the training sample is presented. The dots on the graph indicate the points obtained during the experiment. The dotted line is a standard approximation approach. The solid line is its interpolation modification. It can be seen that on average, as expected, the relative approximation error on small training samples for the interpolation modification of the approach is several times smaller.

To further confirm the hypothesis that "distant" boundaries must be taken into account when approximating weights on a small training sample, we will conduct another experiment.

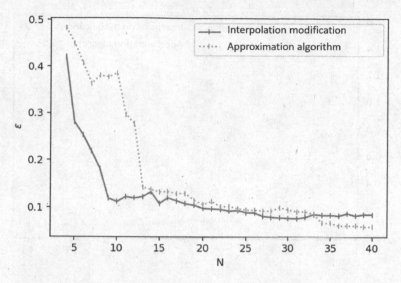

Fig. 2. Dependence of the approximation error on the size N of the training sample

Artificially limit the number of interpolation nodes (and hence the degree of the interpolation polynomial) to 6: we will take into account only the three largest values of $X(l,j)$ and the three smallest $Y(l,j)$. That is, "distant" boundaries are not taken into account in this experiment. The results of the calculations are shown in the graph below.

Finally on Fig. 3 the dependence of the approximation error ε on the size N of the training sample is presented when the degree of the interpolation polynomial is limited. It can be seen that in comparison with the experiment, the results of which are shown in Fig. 2, the approximation error has increased at all points of the graph. Thus, the hypothesis about the need to take into account "distant" boundaries when approximating the weight coefficients is confirmed.

On Fig. 4 we present the results of an experiment using cubic spline interpolation. The results of this experiment are comparable to the results obtained using the Lagrange interpolation polynomial. Therefore, the authors stop at choosing the formula (11) as an expression for approximating the weighting coefficients of the problem.

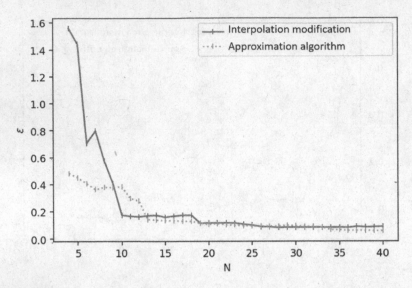

Fig. 3. Dependence of the approximation error on the sample size under the polynomial degree restriction

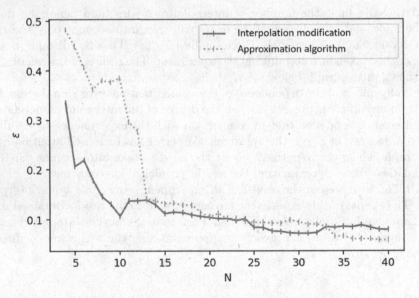

Fig. 4. Dependence of approximation error on sample size for cubic spline interpolation

8   Conclusion

In the paper we study the linear approximation for a single machine scheduling problem is continued [2]. The objective function linear with respect to the completion times of jobs. It is the total weighted completion times of the jobs. Schedules construct "manually" are optimal with respect to this objective function. Unknown values of the weighting coefficients of the objective function are approximated, which reduces, as shown, to solving a system of linear inequalities with respect to these coefficients. Based on linear approximation, a new approximation-interpolation approach has been developed for the problem of minimizing the total weighted completion times of jobs on the single machine. There is a single machine and many jobs that need to be processed on the machine. Processing times are set for each job. There are no precedence relationships of machine between jobs (there are no constraints on the priority of jobs). Interruptions in jobs processing are prohibited. The schedule is uniquely set by the order of jobs. In the problem of minimizing the total weighted completion times on single machine, it is necessary to find a schedule with the minimum value of the objective function. The idea of the approximation-interpolation approach is to solve an effective system of inequalities and interpolate the resulting boundary values. The result of the algorithm is a set of such weighting coefficients that for each of the given instances, the optimal schedule found for the approximated values of the weighting coefficients with its given optimal schedule corresponding to an unknown true set of weighting coefficients. Thus, for a given generation of input data, the interpolation approach shows a more efficient result than the approximation approach. In particular, it was found that it makes sense to use an interpolation approach [7] to boost approximation for a small set of applications, that is, when we don't have enough information for the approximation approach to work correctly.

In the future, it is planned to study not only the linear approximation of an unknown functions, but also the use of more difficult functions that reflect the features of the problem. In addition, it is planned to study the applicability of these methods to solve practical problems, such as the problem of medical care and the problem of railway planning.

Appendix

Algorithm 1. Pseudocode of partitioning into 2 sets

1: generate j weights
2: generate j durations of work
3: **for** N optimal instances **do**
4:     **for** j weight coefficients and processing times **do**
5:         construct a system of inequalities
6:     **end for**
7: **end for**
8: let L be the set of indices k corresponding to the given pairs of problem instances and their optimal schedules
9: divide L into 2 subsets depending on the location of the works i, j
10: distribute the composed system of inequalities over these two subsets
11: let i and j some arbitrary jobs
12: **for** i **do**
13:     **for** j **do**
14:         **if** i **then** != j
15:             **if** the ratio of weight coefficients is greater than the ratio of processing times **then**
16:                 add jobs to the first subset
17:             **else**
18:                 add jobs to the second subset
19:             **end if**
20:         **end if**
21:     **end for**
22: **end for**

Algorithm 2. Pseudocode of calculation of weighting coefficients

---

1: construct a matrix for the minimum value of the ratio of processing times for the first subset
2: construct a matrix for the minimum value of the ratio of processing times for the second subset
3: **for** the first subset **do**
4:     **for** i **do**
5:         **for** j **do**
6:             **if** i != j **then**
7:                 recalculate the minimal values in the matrix according to the expression
8:             **end if**
9:         **end for**
10:     **end for**
11: **end for**
12: **for** the second subset **do**
13:     **for** i **do**
14:         **for** j **do**
15:             **if** i != j **then**
16:                 recalculate the maximum values in the matrix according to the expression
17:             **end if**
18:         **end for**
19:     **end for**
20: **end for**
21: calculate the unknown weight coefficients

References

1. Lenstra, J., Rinnooy Kan, A., Brucker, P.: Complexity of machine scheduling problems. Discrete Math. **1**, 343–362 (1977)
2. Lazarev, A.A., Pravdivets, N.A., Barashov, E.B.: Approximation of the objective function of scheduling theory problems. In: Abstracts of the 13th International Conference "Intellectualization of Information Processing" (Moscow, 2020), pp. 404–409. Russian Academy of Sciences, Moscow (2020). (in Russian)
3. Meng, D., et al.: Reliability-based optimisation for offshore structures using saddle-point approximation. In: Proceedings of the Institution of Civil Engineers-Maritime Engineering, vol. 173(2), pp. 33–42. Thomas Telford Ltd. (2020)
4. Headley, D., et al.: Approximating the quantum approximate optimisation algorithm. arXiv preprint arXiv:2002.12215 (2020)
5. Huotari, J., et al.: Convex optimisation model for ship speed profile: optimisation under fixed schedule. J. Marine Sci. Eng. **9**(7), 730 (2021)
6. Lazarev, A.A.: Scheduling theory. Methods and algorithms. M.: ICS RAS (2019). (in Russian)
7. Lazarev, A.A., Lemtyuzhnikova, D.V., Tyunyatkin, A.A.: Metric interpolation for the problem of minimizing the maximum lateness for a single machine. Autom. Remote Control **82**(10), 1706–1719 (2021). https://doi.org/10.1134/S0005117921100088
8. Tanaev, V.S., Gordon, V.S., Shfransky, Y.M.: Scheduling Theory: Single-Stage Systems. Kluwer Academic Publishers, Dordrecht (1994)
9. Brucker, P.: Scheduling algorithms. J. Oper. Res. Soc. **50**, 774–774 (1999)
10. Tschernikow, S.: A generalization of Kronecker-Capelli's theorem on systems of linear equations. Matematicheskii Sbornik. **57**(3), 437–448 (1944). (in Russian)
11. Smith, W.E.: Various optimizers for single-stage production. Naval Res. Logist. Q. **3**, 59–66 (1956)

Robot Workspace Approximation with Modified Bicenetred Krawczyk Method

Artem Maminov[1,2]([⊠]) [iD] and Mikhail Posypkin[1,2] [iD]

[1] Federal Research Center "Computer Science and Control" of the Russian Academy of Sciences (FRC CSC RAS), 44/2, Vavilova Street, 119333 Moscow, Russian Federation
artem_maminov@mail.ru

[2] HSE University, 20 Myasnitskaya Ulitsa, 101000 Moscow, Russian Federation

Abstract. The article considers the application of numerical method based on bicentered Krawczyk operator for solving the problem of the robot workspace approximation with box constraints. We applied several modifications for approximation of the solution sets of the indeterminate system of nonlinear equations and compare it with basic method. All methods were tested on a passive orthosis robot, which is part of the lower limb rehabilitation system. A mathematical model of the mechanism kinematics is presented. We evaluate the efficiency of the considered approaches, compute and visualize the robot workspace for different parameters sets.

Keywords: Robot workspace · Rehabilitation system · Undetermined system of the equations · Krawczyk method · Interval analysis

1 Introduction

Since the development of electrical engineering in the middle of the 20-th century, robots of various designs have entered into many areas of our life: conveyor production, construction, aerospace industry, medicine and etc. The information about workspace area of the mechanism is very significant in associated problems: robot's construction optimization, path planning, acceptable area for another mechanism workspace. Usually, the robot is set by the kinematic equations. Kinematic system is the system of the equations, which determines the position of the working tool.

Typically, such systems are indeterminate, since robots have several degrees of freedom. For solving this problem we use the interval analysis methods. These methods allow to find reliable solution sets, what we use for workspace approximation of the passive orthosis in the lower limb rehabilitation system.

This research was supported by the Ministry of Science and Higher Education of the Russian Federation, project No 075-15-2020-799.

N. Olenev et al. (Eds.): OPTIMA 2022, LNCS 13781, pp. 238–249, 2022.
https://doi.org/10.1007/978-3-031-22543-7_17

2   Related Works

Basically this work consists of two research areas: interval analysis and approximation of the robotic mechanism's workspace. Interval analysis (interval mathematics, interval computation) is a method developed in the 50s and 60s of the last century. It's used for determination of the rounding errors boundaries and measurement errors in mathematical calculations. Thus numerical methods that give reliable results using intervals for calculations have been developed.

One of the first books, where the interval analysis was mentioned, was the book [4]. The author proposes arithmetic operations on range numbers to improve the reliability of digital systems. The birth of modern interval arithmetic corresponds to the appearance of the book [14]. Such methods, which has been mentioned and described in these books, operates not with specific values, but with ranges of possible values specified by the beginning and end of the range.

The development of interval analysis is also connected with Arnold Neumaier and Eldon Hansen's works. Professor Neumaier continued the research of the application of interval methods to the different mathematical problems [16]. Eldon Hansen also contributed to the study of interval techniques and found out one of the most widely used methods for solving systems of equations [5].

Nowadays the most famous researcher in the sphere of interval analysis in CIS is professor Sergey Shariy, who follows the base of Novosibirsk mathematical school. He studied different applications of interval technique for the solving of the systems of equations, optimization problem and sets approximation [17,18]. He also proposed the usage of Baumann theorem [1] to the Krawczyk operator [9] for solving undetermined systems of nonlinear equations. This method can be called bicenterd Krawczyk operator and exactly this method is used in this work.

One of the most fundamental work, especially in the sphere of parallel robots, is the book [13]. It contains basic approximation methods and detailed information about parallel robots. Other researchers study concrete robotic system and its workspace. For example in [10] the Passive Orthosis mechanism is analysed and its workspace bounded with non-uniform covering method. But such type of methods is more complex than ours and the authors have to extract inequalities from initial system, which are used in algorithms. In [3] the scientists use geometrical approach for workspace determination, but such approaches can't be used for all robots, since system can be much more complicated and it takes a lot of time to analyze geometrical characteristics.

The most relevant work is the article [2]. Hansen-Sengupta method is tested for different robots workspace approximation. The branch-and-prune technique, which is commonly used in interval analysis algorithms is also used. In our work we propose new interval technique based on Baumann bicentered theorem. This algorithm is poorly studied and tested only on synthetic system with simple structure [11]. We assume that this method can be more precise than other classical methods.

3   Problem Statement

The conceptual design of the lower limb rehabilitation system is shown in Fig. 1. The structure of the rehabilitation system used an active 3-PRRR parallel robot proposed by Kong and Gosselin [8] and a passive orthosis. The conceptual passive orthosis design based on a serial RRRR chain is shown in (Fig. 2).

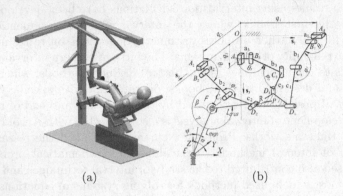

(a)   (b)

Fig. 1. Lower limb rehabilitation system: a — 3D-model of the system; b — design scheme

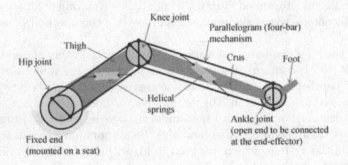

Fig. 2. Conceptual design of the proposed passive orthosis system

In this work we use 2-D model of the proposed system:

$$\begin{cases} x_P = l_a \cos(\alpha) + l_b \cos(\alpha + \beta), \\ y_P = l_a \sin(\alpha) + l_b \sin(\alpha + \beta), \end{cases} \tag{1}$$

where l_a — is the length of thigh bar, l_b — is the length of the crus bar, α — is the angle between hip joint and OX axes, β — is the angle between thigh and crus bars.

Thus, we consider the problem of solving undetermined system of nonlinear equations. Let us rewrite in more appropriate form:

$$F(u, v) = 0, \tag{2}$$

where u, v are vectors in the spaces $\mathbb{R}^m$, $\mathbb{R}^n$, respectively, and $F : \mathbb{R}^{m+n} \to \mathbb{R}^n$ is a continuously differentiable mapping. Given a box $U = [\underline{u}_1, \overline{u}_1] \times \cdots \times [\underline{u}_m, \overline{u}_m] \subseteq \mathbb{R}^m$ and a box $V = [\underline{v}_1, \overline{v}_1] \times \cdots \times [\underline{v}_n, \overline{v}_n] \subseteq \mathbb{R}^n$, let us define the *solution set* Ω as

$$\Omega = \{u \in U \subseteq \mathbb{R}^m \mid \exists v \in V \subseteq \mathbb{R}^n \text{ such that } F(u, v) = 0\}, \tag{3}$$

i.e. as the set formed by all such parameters $u \in U$ that there exists a solution v for the equations system $F(u, v) = 0$.

For our problem we have box $U = [\underline{u}_1, \overline{u}_1] \times [\underline{u}_2, \overline{u}_2]$, which limits the workspace of the mechanism. The box $V = [\underline{v}_1, \overline{v}_1] \times [\underline{v}_2, \overline{v}_2]$ limits the angles α and β accordingly:

$$F(u, v) = \begin{cases} u_1 - l_a \cos(v_1) - l_b \cos(v_1 + v_2), \\ u_2 - l_a \sin(v_1) - l_b \sin(v_2 + v_2). \end{cases} \tag{4}$$

The robot workspace consists of all points, which satisfies Eq. 4 and in fact it's all points, which can be achieved by working tool. In our program complex we construct approximation of such points by the sets of inner boxes (S_{in}) and border boxes (S_b) such that

$$\cup_{u \in S_{in}} \subseteq \Omega \subseteq \cup_{u \in S_{in} \cup S_b} \subseteq U. \tag{5}$$

4 Preliminaries

All algorithms presented in the article use the interval analysis techniques [6, 15, 16], in particular, methods for constructing reliable enclosures for the ranges of functions. Our notation follows the informal standard [7].

Interval $u \subseteq \mathbb{R}$ is a set of real numbers enclosed between its left and right endpoints, that is, $u = \{u \in \mathbb{R} \mid \underline{u} \le u \le \overline{u}\}$. The set of all intervals is denoted as $\mathbb{IR}$. An n-dimensional box is defined as the direct Cartesian product of n intervals, and the set of such boxes is denoted by $\mathbb{IR}^n$. In our article, we use boldface font to denote intervals and boxes, while normal font denotes real numbers and vectors. The diameter of a box x is defined in the usual way, as the maximum distance between points in the box:

$$\text{diam}(x) = \sup \{\|a - b\| \mid a, b \in x\} = \|\overline{x} - \underline{x}\|.$$

The scale function is used to expand input interval:

$$\text{scale}(x, c) = [\underline{x} - \underline{x} \cdot c, \overline{x} + \overline{x} \cdot c]$$

Arithmetic operations can be defined between intervals, and thus we get various interval arithmetics. In our work, we use the classical interval arithmetic, an algebraic system formed by the intervals so that the result of any arithmetic operation "$\star$" between the intervals is defined "by representatives", as

$$\boldsymbol{x} \star \boldsymbol{y} = \{ \, x \star y \mid x \in \boldsymbol{x}, y \in \boldsymbol{y} \, \} \qquad \text{for each operation} \quad \star \in \{ +, -, \cdot, / \}.$$

Expanded constructive formulas for interval arithmetic operations are as follows (see e.g. [6,15,16]):

$$\boldsymbol{x} + \boldsymbol{y} = \left[\underline{x} + \underline{y}, \; \overline{x} + \overline{y} \right], \qquad \boldsymbol{x} - \boldsymbol{y} = \left[\underline{x} - \overline{y}, \; \overline{x} - \underline{y} \right],$$

$$\boldsymbol{x} \cdot \boldsymbol{y} = \left[\min\{\underline{x}\,\underline{y}, \underline{x}\,\overline{y}, \overline{x}\,\underline{y}, \overline{x}\,\overline{y}\}, \; \max\{\underline{x}\,\underline{y}, \underline{x}\,\overline{y}, \overline{x}\,\underline{y}, \overline{x}\,\overline{y}\} \right],$$

$$\boldsymbol{x}/\boldsymbol{y} = \boldsymbol{x} \cdot \left[1/\overline{y}, \; 1/\underline{y} \right] \qquad \text{for} \quad \boldsymbol{y} \not\ni 0.$$

For basic functions usage (such as power function, logarithmic function) in interval calculations, we should redefine them using the monotonicity properties of the function. For monotone functions of one variable, the range of values is determined quite simply. If $f : \mathbb{R} \to \mathbb{R}$ is monotone on the interval $\boldsymbol{a} = \left[\underline{a}, \overline{a}\right]$, then the range of values of the function on it can be obtained by calculating the value of the function at the ends of the interval. In the case of a monotonically increasing function: $f(\boldsymbol{a}) = \left[f(\underline{a}), f(\overline{a})\right]$. If function is monotonically decreasing: $f(\boldsymbol{a}) = \left[f(\overline{a}), f(\underline{a})\right]$. The trigonometric sine and cosine functions are not monotonic for the entire number axis. Thus, we need to check if the interval contains minimum and maximum values of the sine and cosine function. Obviously, such points are $-\frac{\pi}{2}, \frac{\pi}{2}$ and $0, \pi$ for the sine and cosine functions respectively.

Algorithm 1. Sine algorithm

$\quad \boldsymbol{x}$ — input interval

1: $\boldsymbol{y} = [\underline{y}, \overline{y}] = [\sin(\underline{x}), \sin(\overline{x})]$

2: **if** $\left\lceil \frac{\underline{x} - \frac{\pi}{2}}{2\pi} \right\rceil \leq \left\lfloor \frac{\overline{x} - \frac{\pi}{2}}{2\pi} \right\rfloor$ **then**

3: $\quad\quad b = 1$

4: **else**

5: $\quad\quad b = \max \boldsymbol{y}$

6: **end if**

7: **if** $\left\lceil \frac{\underline{x} + \frac{\pi}{2}}{2\pi} \right\rceil \leq \left\lfloor \frac{\overline{x} + \frac{\pi}{2}}{2\pi} \right\rfloor$ **then**

8: $\quad\quad a = -1$

9: **else**

10: $\quad\quad a = \min \boldsymbol{y}$

11: **end if**

12: **return** Interval$[a, b]$

Algorithm 1 is a realization of the interval sine function. In lines 2 and 7 we check if critical points $(-\frac{\pi}{2}, \frac{\pi}{2})$ are inside of the input interval $\boldsymbol{x}$. If we find them, we can conclude, that minimum and maximum values for the resulting interval are in –1 and 1 accordingly. Otherwise, we record left or right border of the interval as $\min([\sin(\underline{x}), \sin(\overline{x})])$ or $\max([\sin(\underline{x}), \sin(\overline{x})])$ accordingly. The cosine function calculation is similar to this and differs only in critical points.

An *interval extension* of a function $f : \mathbb{R}^n \to \mathbb{R}^m$ is an interval function $f : \mathbb{IR}^n \to \mathbb{IR}^m$ such that

1) $f(x) = f(x)$ for any $x \in \mathbb{R}^m$ from the domain of f,
2) $f(x)$ is inclusion monotonic, i.e., for any $x, y \in \mathbb{IR}^n$, $x \subseteq y$ implies $f(x) \subseteq f(y)$.

The simplest way to obtain an interval extension is called *natural interval extension*. In this approach, all occurrences of variables in the expression for a function are replaced with intervals. The interval enclosure is evaluated by executing all operations according to interval arithmetic rules. We denote the natural interval extension of a function f as $\natural f$.

4.1 Recurrent Forms

The equivalent recurrent forms are widely used for the proofing of the existence of the solution. We consider the System 2.

Using Brouwer Fixed-Point theorem we can prove the inclusion $u \subseteq \Omega$. We can rewrite system to the equivalent recurrent form:

$$v = G(u, v), \tag{6}$$

where $G(u, v) = v - \Lambda F(u, v)$ and Λ is a nonsingular $n \times n$ matrix.

4.2 Basic Krawczyk Method

The further presented algorithms relies on interval extension $G(u, v)$. In this work two such interval extensions presented: Krawczyk method (we call it "basic", since it uses the classical Krawczyk operator) and bicentered Krawczyk method, which is the modernization of basic Krawczyk operator. In fact $G(u, v)$ is a vector valued function

$$G(u, v) = \begin{pmatrix} g_1(u, v) \\ \vdots \\ g_n(u, v) \end{pmatrix},$$

and we can construct an interval extension for each component function $g_i(u, v)$. . Specifically, we use the so-called differential centered form (also known as *mean-value form*, see [15, 16]):

$$\tilde{g}_i(u, v, c) = \natural g_i(u, c) + \natural g_i'(u, v)^\top (v - c), \tag{7}$$

where $c \in v$ is a point from the interval v and a function

$$g_i'(u, v) = \begin{pmatrix} \frac{\partial g_i(u,v)}{v_1} \\ \vdots \\ \frac{\partial g_i(u,v)}{v_n} \end{pmatrix}.$$

is the gradient of the mapping g_i with the respect to variables from v. The natural interval extension $\natural$ is applied to vectors in element-wise manner.

Taking $c = \mathrm{mid}\, v$, we obtain the classical Krawczyk operator:

$$K(u,v) = \begin{pmatrix} \tilde{g}_1(u,v,\mathrm{mid}\,v) \\ \vdots \\ \tilde{g}_n(u,v,\mathrm{mid}\,v) \end{pmatrix}. \tag{8}$$

This operator is an interval extension of the mapping $G(u,v)$ [19], i.e. $G(u,v) \subseteq K(u,v)$.

4.3 Bicentered Method

The modification of Krawczyk operator based on the Baumann theorem [1] can give tighter bounds. Such modification was proposed in [19]. Denote $d^{(i)} = \natural\, g_i'(u,v)$, $d^{(i)} \in \mathbb{IR}^n$. According to the Baumann theorem,

$$g_i(u,v) \subseteq \tilde{g}_i\left(u,v,\hat{c}^{(i)}\right) \cap \tilde{g}_i\left(u,v,\check{c}^{(i)}\right), \tag{9}$$

where for $j = 1, 2, \ldots, n$

$$\hat{c}_j^{(i)} = \begin{cases} \underline{v}_j, & \text{if } \overline{d}_j^{(i)} \le 0, \\ \overline{v}_j, & \text{if } \underline{d}_j^{(i)} \ge 0, \\ \dfrac{\underline{d}_j^{(i)}\,\underline{v}_j - \overline{d}_j^{(i)}\,\overline{v}_j}{\underline{d}_j^{(i)} - \overline{d}_j^{(i)}}, & \text{if } \underline{d}_j^{(i)} < 0 < \overline{d}_j^{(i)}, \end{cases} \qquad \check{c}_j^{(i)} = \begin{cases} \overline{v}_j, & \text{if } \overline{d}_j^{(i)} \le 0, \\ \underline{v}_j, & \text{if } \underline{d}_j^{(i)} \ge 0, \\ \dfrac{\overline{d}_j^{(i)}\,\underline{v}_j - \underline{d}_j^{(i)}\,\overline{v}_j}{\overline{d}_j^{(i)} - \underline{d}_j^{(i)}}, & \text{if } \underline{d}_j^{(i)} < 0 < \overline{d}_j^{(i)}. \end{cases}$$

We define the *bicentered Krawczyk operator* $K_{bic}(u,v)$ as follows:

$$K_{bic}(u,v) = \begin{pmatrix} \tilde{g}_1\left(u,v,\hat{c}^{(1)}\right) \cap \tilde{g}_1\left(u,v,\check{c}^{(1)}\right) \\ \vdots \\ \tilde{g}_n\left(u,v,\hat{c}^{(n)}\right) \cap \tilde{g}_n\left(u,v,\check{c}^{(n)}\right) \end{pmatrix}, \tag{10}$$

where $g_i\left(u,v,c^{(i)}\right)$ is the classical Krawczyk operator. Similar to the Krawczyk operator, this operator is also an interval extension of the mapping $G(u,v)$ [19], i.e. $G(u,v) \subseteq K_{bic}(u,v)$.

5 Algorithms Description

At line 1 of the Algorithm 2, the box U is partitioned into equal smaller boxes on a uniform grid with k nodes in every dimension, so that $U = \cup_{i=1}^{i=m^k} U^{(i)}$. The computed boxes are put in a list L. For each element of the list $L = \left\{ U^{(i)} \right\}$ the procedure checkBox is called (lines 3–4). If the procedure returns IN or $INDET$ this box is placed to list S_{in} or list S_b respectively. We also introduce new approach of using input box V. Usually, in common algorithms we take the hole box, but the recurrent algorithms can't work successfully with all variables

---

Algorithm 2. Cover algorithm with enlargement input box

U, V — bounding boxes for u and v

1: Construct a uniform meshes $L_u = \left\{U^{(i)}\right\}$, $U = \cup_{i=1}^{i=m^k} U^{(i)}$ and $L_v = \left\{V^{(i)}\right\}$, $V = \cup_{i=1}^{i=m^n} V^{(i)}$

2: Initialize lists $S_{in} = \{\}, S_b = \{\}$

3: **for all** $u \in L_u$ **do**

4:     $check = OUT$

5:     **for all** $v \in L_v$ **do**

6:         $r := \mathtt{checkBox}(u, v)$

7:         **if** $r = IN$ **then**

8:             $S_{in} := S_{in} \cup u$

9:             $check = OUT$

10:            **break**

11:        **else if** $r = INDET$ **then**

12:            $check = INDET$

13:        **end if**

14:    **end for**

15:    **if** $check = INDET$ **then**

16:        $S_b := S_b \cup u$

17:    **end if**

18: **end for**

19: **return** S_{in}, S_b

(for example angles). That's why we also construct an uniform grid L_v based on input box V. The iteration throw this list added to the algorithm (line 5). The smaller boxes v are checked, which leads to more precise evaluation and better estimation of the inside boxes. The theoretical aspect of such method will be explained after the Algorithm $\mathtt{checkBox}$.

Lists S_{in} and S_b constructed by the algorithm satisfy the following condition: $\cup_{u \in S_{in}} \subseteq \Omega \subseteq \cup_{u \in S_{in} \cup S_b} \subseteq U$ providing that procedure $\mathtt{checkBox}$ correctly classifies the box, where Ω is the solution set: $\Omega = \{u \in U \subseteq \mathbb{R}^m \mid \exists v \in V \subseteq \mathbb{R}^n$ such that $F(u, v) = 0\}$. However the algorithm does not guarantee the tightness of the constructed approximation. The collection S_b can contain boxes having no points from the boundary of a set Ω. Let us consider $\mathtt{checkBox}$ in detail.

In line 1 of the Algorithm 3, we compute an interval extension $G(u, v)$, which results in a box v'. Then (lines 2–4) we use *inflation* technique [12] starting from 6 iteration. This method can improve the determination of the solution set. In lines 6–7 we check the fulfillment of the inclusion $v' \in v$, which leads to $u \in \Omega$ according to Brouwer Fixed-Point Theorem. Thus if the answer is positive, the box u is classified as *internal*. Otherwise, the intersection $v'' = V \cap v$ is computed in case of enlarging method. If it is empty, then the box u lies outside the solution set Ω. Otherwise, in line 12, we check that the box was notably changed during the iteration and the number of the iterations is lower than maximum iterations value. If this is not true, the box is classified as a boundary. If the change is large, the procedure is called recursively (line 15) with a changed box v''. Eventually,

---

Algorithm 3. Check box

u, V, v — boxes, $k = 0$ — number of the iteration
1: $v' := G(u, v)$
2: if k>5 then
3:     $v' = \text{scale}(v', 1.1)$
4:     $v' = v' \cap V$
5: end if
6: if $v' \subseteq v$ then
7:     return IN
8: else
9:     $v'' = v' \cap V$
10:    if $v'' = \varnothing$ then
11:        return OUT
12:    else if $|\text{diam}(v) - \text{diam}(v'')| \leq \varepsilon$ or $k = max\_iter$ then
13:        return $INDET$
14:    else
15:        return $\text{checkBox}(u, v'')$
16:    end if
17: end if

---

the procedure terminates due to a little progress in the box reduction or because the conditions in lines 6 or 10 are satisfied.

6   Experimental Results

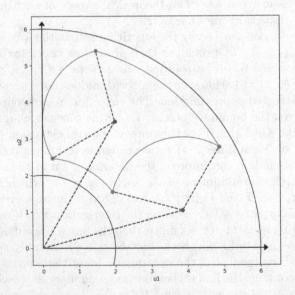

Fig. 3. Analytical workspace determination for $v_1 = [\frac{\pi}{12}, \frac{\pi}{6}]$, $v_2 = [\frac{\pi}{4}, \frac{5\pi}{6}]$. Red figure is the theoretical workspace, blue lines are limits for the considered parameters.

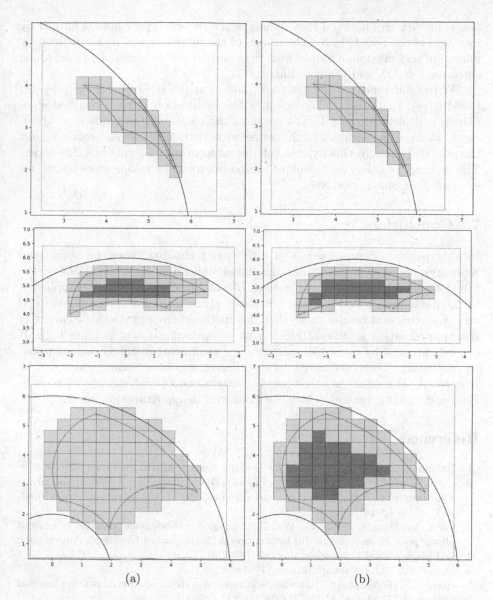

Fig. 4. The results produced by bicentered Krawczyk method without (a) and with (b) enlarging technique for uniform grid of size 15 × 15 for rehabilitation system

For the computational experiments we use System 4. We use constant parameters $l_a = 4$ and $l_b = 2$ and different initial angles for the system, since the whole workspace is presented as ring with circles of radii of $l_a - l_b$ and $l_a + l_b$. The analytical workspace is constructed as area between four points (*points*), which are produced by calculating u_1 and u_2 from cartesian product of $v_1 \times v_2$. We also construct arcs between these points, since the workspace area is formed by rotations of the bars (Fig. 3). The ranges of the intput u_1 and u_2 are defined by minimum and maximum values from first and second columns of *points*, which correspond to OX and OY coordinates.

We test different variations of the v_1 and v_2 angles $(([\frac{\pi}{12}, \frac{\pi}{6}], [\frac{\pi}{12}, \frac{\pi}{3}]); ([\frac{\pi}{4}, \frac{\pi}{2}], [\frac{\pi}{4}, \frac{\pi}{2}]); ([\frac{\pi}{12}, \frac{\pi}{6}], [\frac{\pi}{4}, \frac{5\pi}{6}]))$. The result of the application of the described algorithms is depicted in Fig. 4. The usage of the enlarging technique can significantly increase the accuracy of the inside boxes determination and removing out the outside boxes. In this experiment the uniform grid L_v contains 225 nodes. The accuracy of enlarging technique can be improved by taking more nodes (for example for Combination №1).

7   Conclusions

In this paper, we presented a modification technique based on bicentered Krawczyk operator for determination inner and outer approximations of the solution sets for indeterminate systems of non-linear equations. We applied it to the mechanism based on passive orthosis of the lower limb rehabiliation system. Experimental results show, that this method can improve the accuracy of the approximation significantly. For now computational cost is quite high. It can be improved by using high performance computing methods such as parallel computing. There are also a lot of different robotic systems, which can be researched. The usage of other interval extensions can be considered such as Hansen-Sengupta method, which can improve the approximation accuracy.

References

1. Baumann, E.: Optimal centered forms. BIT Numer. Math. **28**(1), 80–87 (1988)
2. Caro, S., Chablat, D., Goldsztejn, A., Ishii, D., Jermann, C.: A branch and prune algorithm for the computation of generalized aspects of parallel robots. Artif. Intell. **211**, 34–50 (2014)
3. Chen, Y., Han, X., Gao, F., Wei, Z., Zhang, Y.: Workspace analysis of a 2-dof planar parallel mechanism. In: International Conference of Electrical, Automation and Mechanical Engineering, pp. 192–195 (2015)
4. Dwyer, P.S.: Linear computations (1951)
5. Hansen, E., Sengupta, S.: Bounding solutions of systems of equations using interval analysis. BIT Numer. Math. **21**(2), 203–211 (1981)
6. Kearfott, R.B.: Rigorous global search: continuous problems. Springer Science & Business Media, Heidelberg (2013). https://doi.org/10.1007/978-1-4757-2495-0
7. Kearfott, R.B., Nakao, M.T., Neumaier, A., Rump, S.M., Shary, S.P., van Hentenryck, P.: Standardized notation in interval analysis. Comput. Technol. **15**(1), 7–13 (2010)
8. Kong, X., Gosselin, C.M.: Kinematics and singularity analysis of a novel type of 3-CRR 3-DOF translational parallel manipulator. Int. J. Robot. Res. **21**(9), 791–798 (2002)
9. Krawczyk, R.: Interval extensions and interval iterations. Computing **24**(2–3), 119–129 (1980)

10. Malyshev, D., Nozdracheva, A., Dubrovin, G., Rybak, L., Mohan, S.: A numerical method for determining the workspace of a passive orthosis based on the RRRR mechanism in the lower limb rehabilitation system. In: Pisla, D., Corves, B., Vaida, C. (eds.) EuCoMeS 2020. MMS, vol. 89, pp. 138–145. Springer, Cham (2020). https://doi.org/10.1007/978-3-030-55061-5_17

11. Maminov, A.D., Posypkin, M.A., Shary, S.P.: Reliable bounding of the implicitly defined sets with applications to robotics. Procedia Comput. Sci. **186**, 227–234 (2021)

12. Mayer, G.: Epsilon-inflation in verification algorithms. J. Comput. Appl. Math. **60**(1–2), 147–169 (1995)

13. Merlet, J.P.: Parallel robots, vol. 128. Springer Science & Business Media, Dordrecht (2006). https://doi.org/10.1007/1-4020-4133-0

14. Moore, R.E.: Interval analysis, vol. 4. Prentice-Hall Englewood Cliffs (1966)

15. Moore, R.E., Kearfott, R.B., Cloud, M.J.: Introduction to interval analysis. SIAM (2009)

16. Neumaier, A.: Interval methods for systems of equations. Cambridge University Press, Cambridge (1990)

17. Shary, S.P.: Algebraic approach to the interval linear static identification, tolerance, and control problems, or one more application of kaucher arithmetic. Reliab. Comput. **2**(1), 3–33 (1996)

18. Shary, S.P.: A new technique in systems analysis under interval uncertainty and ambiguity. Reliab. Comput. **8**(5), 321–418 (2002)

19. Shary, S.P.: Krawczyk operator revised. In: Proceedings of International Conference on Computational Mathematics ICCM-2004, Novosibirsk, Russia, June 21–25, 2004, pp. 307–313. Institute of Computational Mathematics and Mathematical Geophysics (ICM&MG) (2004)

Author Index

Printed in the United States
by Baker & Taylor Publisher Services

Printed in the United States
by Baker & Taylor Publisher Services